Rainfed Agriculture and Watershed Management

NIPA® GENX ELECTRONIC RESOURCES & SOLUTIONS P. LTD.
New Delhi-110 034

About the Author

Dr. R.K. Nanwal, Professor (Retd.), Department of Agronomy CCS Haryana Agricultural University, Hisar-125004 (India). For being M.Sc (Ag.) in Agronomy from HAU, Hisar in 1981, he was awarded Ph.D. in Agronomy by IARI, N.Delhi in 1992. He did Post Doctorate specialized training in water resource management in Rainfed Agriculture at University of Georgia, USA. Attended many technical trainings, refresher courses and summer institutes in various aspects of agriculture for updating the knowledge. Eight such trainings were attended in India and one in USA. Acted as course coordinator/Director of various courses organized by counseling and placement centre/AAREM/Department of Agronomy etc. Designed 40 such courses. He has teaching experience of more than 35 years in the university. Developed new courses on Farming Systems, Sustainable Agriculture and Rainfed Agriculture. Guided 7 M.Sc and 8 Ph.D students as Major Advisor. He is member of advisory committee in number of post graduate students. He has published 9 books, 10 manuals, 7 book chapters, 153 full research papers, 75 general articles, 35 research notes and presented 60 papers in various national/ international seminars/conferences. Received "Indian Council of Agricultural Research, N. Delhi Award" for outstanding Multidisciplinary team research in agriculture and allied sciences for the triennium 1994-96, awarded on July 16, 1997 by Union Minister of Agriculture, Govt of India. Awarded CCS Haryana Agricultural University "Best Teacher Award" (Indian Council of Agricultural Research sponsored) for the year 2005-2006. He was awarded "Rashtriya Gaurav Award" by India International Friendship Society on Nov.11, 2011. He was also awarded Best Citizens of India Award by International Publishing House, N. Delhi on Dec. 28, 2011.

Rainfed Agriculture and Watershed Management

R.K. Nanwal
Professor (Retd.)
Department of Agronomy
CCS Haryana Agricultural University
Hisar 125 004, Haryana, India

NIPA® GENX ELECTRONIC RESOURCES & SOLUTIONS P. LTD.
New Delhi-110 034

NIPA® GENX ELECTRONIC
RESOURCES & SOLUTIONS P. LTD.

101,103, Vikas Surya Plaza, CU Block
L.S.C. Market, Pitam Pura, New Delhi-110 034
Ph : +91 11 27341616, 27341717, 27341718
E-mail: newindiapublishingagency@gmail.com
web: www.nipabooks.com

For customer assistance, please contact
Phone: + 91-11-27 34 17 17
Fax: + 91-11- 27 34 16 16
E-Mail: feedbacks@nipabooks.com

ISBN: 978-81-19254-04-0

Composed and Designed by NIPA®.

Dedicated to my mother

SMT. SHANTI DEVI

Always a source of inspiration to me

Preface

In India, rain dependent areas are vast and they have a great contribution to make in agricultural production, with just 1/40th of the world land. Of the 141 M ha of net sown area in the country, 80 M ha is rainfed. Rainfed agriculture contributes 40% of food grain production and supports 2/3^{rds} of the livestock population. Despite considerable progress in irrigation development over the five year plans, 85% of coarse cereals, 83% in pulses, 42% in rice, 70% in oilseeds and 65% in cotton are still cultivated as rainfed. Most of the rainfed lands are typified by highly fragile natural resource base; the rainfall is low; soils are often coarse textured, sandy, inherently low in fertility, organic matter and water holding capacity; and are easily susceptible to wind and water erosion. Deterioration of natural resources is the main issue threatening sustainable development of rainfed agriculture, more so in the Third World countries. The need for comprehensive information on rainfed agriculture and watershed management for students of agriculture has been felt for quite some time. I hope this book will prove a fillip.

The topics covered are most relevant in view of growing interests in rainfed agriculture technologies. The broad chapters included in the book are: Rainfed agriculture: Introduction, types, characteristics, distribution and importance; History of rainfed agriculture & watershed in India; Problems and prospects of rainfed agriculture in India ; Soil and climatic conditions prevalent in rainfed areas; Drought: types, effect of water deficit on physio- morphological characteristics of the plants, Mechanism of crop adaptation under moisture deficit condition; Water harvesting: importance, its techniques, Efficient utilization of water through soil and crop management practices, Study of mulches and antitranspirants; Management of crops in rainfed areas; Contingent crop planning for aberrant weather conditions; Concept, objective, principles and components of watershed management, factors affecting watershed management; Integrated watershed development.

I acknowledge with indebtedness to authors of books/literature from which some of the material in this book has been drawn. Inspite of the best efforts, it is possible that some errors may have crept into the compilation. The readers are requested to kindly let me know the mistakes so that these can be taken care of in the further edition.

I am thankful to my family especially my wife Mrs. Raj Rani for supporting me during the course of writing the book. Finally, I thank the publisher for bringing out this book so efficiently and promptly.

Author

Contents

Section-II: Practical Exercises

Section-I: Theory

Chapter 1

Rainfed Farming : Introduction Types, Characteristics, Distribution and Importance

Importance of Rainfed Farming

In India, rain dependent areas are vast and they have a great contribution to make in agricultural production, with just 1/40th of the world land. Of the 141 M ha of net sown area in the country, 80 M ha is rainfed. Rainfed agriculture contributes 40% of food grain production and supports 2/3^{rds} of the livestock population. Despite considerable progress in irrigation development over the five year plans, 85% of coarse cereals, 83% in pulses, 42% in rice, 70% in oilseeds and 65% in cotton are still cultivated as rainfed. A large portion of industrially important crops such as cotton, groundnut and castor are cultivated under dry land conditions and has a great contribution to make to the production of food, fibre, fuel and furniture timber etc. Though, impressive gains were noted in some of the rainfed crops in recent times, the gap between attainable and farmers' yields still remain high which is a major cause of concern. Small and marginal farmers who are the backbone of rainfed farming are resource poor and risk averse. Moreover, both public and private investments in technology adoption and infrastructure have been quite low in rainfed agriculture resulting in a vicious circle of low yields-low net returns-low investments in improved technologies.

With several resource management problems emerging in irrigated regions, rainfed agriculture offers scope to contribute to the growing food needs of future particularly of pulses and edible oils. The high yield gaps between attainable yields realized by progressive farmers using improved technologies and those actually realized by the farmers at large in any given region offer scope for enhancing overall productivity of rainfed crops anywhere between 50 to 70% in the near term with the adoption of the available technologies, provided resource and institutional constraints in delivering the technology and inputs are addressed timely and adequately. With the promised break-through in yields through genetic manipulation and large scale adoption of water saving technologies through conservation agriculture, the yield gains can be further enhanced in the medium to long term.

The 80 million hectares of rainfed farming area is not uniform for resource base, opportunities and constraints. There are areas receiving high rainfall and fertile deep soils where crop yields can be enhanced significantly by double cropping combined with supplemental irrigation with water harvesting. For the relatively lower rainfall regions, the overall system productivity needs to be enhanced by integrating crops, trees and livestock. This agro-eco region based differential approach is the key for realizing the full potential of rainfed agriculture besides addressing the dimension of regional equity.

In addition to the small holders dominance, rainfed agriculture also faces new challenges of climate variability. Besides, changed rainfall patterns, it is predicted that extreme events are likely to increase in the country resulting in more droughts and floods. This poses new challenges in terms of rainwater management and soil conservation. Dryland Agriculture may be classified into three groups on the basis of annual rainfall.

a) Dry farming: Cultivation of crops in areas where annual rainfall is less than 750 mm and crop failures due to prolonged dry spells during crop period are most common. Dry farming is practiced in arid regions with the help of moisture conservation practices. Alternate landuse system is suggested in this region.

b) Dryland farming: Cultivation of crops in areas where annual rainfall is more than 750 mm but less than 1150 mm is called dryland farming. Dry spells may occur, but crop failures are less frequent. Higher Evapotranspiration (ET) than the total precipitation is the main reason for moisture deficit in these areas. The soil and moisture conservation measures are the key for dryland farming practice in semi-arid regions. Drainage facility may be required especially in black soils.

c) Rainfed farming: Cultivation of crops in regions where rainfall is more than 1150 mm. There is less chances of crop failures due to dry spells. There is adequate rainfall and drainage becomes the important problem in rainfed farming. This farming is practiced in humid regions.

Steiner *et al.* (1988) consider aridity index concept of the United Nations conference on desertification based on the balance between precipitations (P) and evapotranspiration (ET) to be appropriate for wide scale adoption. According to this definition areas with P/ET ratio between 0.03 and 0.20 are arid and areas with the ratio between 0.2 to 0.5 are semi arid.

Another influential definition is based on "growing period concept of FAO". Areas having 9 and 74 days are classified as arid and there with a growing period between 75 and 119 days semi arid. The growing period is the number of days during a year, when precipitation exceeds half the potential evapotranspiration, plus a period to use an assumed 100 mm of water form excess (or less, if not available) stored in the soil profile.

Indian Council of Agricultural Research (ICAR) while establishing the dryland centers in different agro climatic zones of the country in 1970 used the simple formula of Thornth Waite for estimating the moisture deficiency index.

$$\text{Moisture Deficiency index} = \frac{\text{P-PE}}{\text{PE}} \times 100$$

Where,

P = Precipitation and

PE = Potential evaporation

According to this classification the limit for semi-arid climates is given by the values of the aridity index varying from -33.3 to 66.7.

Dryland v/s Rainfed farming

S.No.	Constituent	Dryland farming	Rainfed farming
1	Rainfall	< 800 mm	> 800 mm
2	Moisture availability to the crop	Shortage	Enough
3	Growing season (days)	< 200	> 200
4	Growing region	Arid and semi arid areas	Humid & subhumid regions
5	Cropping system	Single crop or intercropping	Intercropping or Double cropping
6	Constraints	Wind and water erosion	Water erosion

Characteristics of rainfed agriculture

Rainfed areas may be characterized by the following features:

- Uncertain, ill-.distributed and limited annual rainfall
- Occurrence of extensive climatic hazards like drought, flood etc.
- Undulating soil surface
- Occurrence of extensive and large holdings
- Practice of extensive agriculture i.e. prevalence of mono cropping etc.
- Relatively large size of fields
- Similarity in types of crops raised by almost all the farmers of a particular region

- Very low crop yield
- Poor market facility for the produce
- Poor economy of the farmers
- Poor health of cattle as well as farmers

Distribution of rainfed farming

Global distribution of rainfed areas

Arid and semi-arid regions comprise almost 40 per cent of the world land area and are inhabitated by some 700 million people. Approximately 60 per cent these drylands is in developing countries. Low rainfall area constitutes from 75–100 per cent of land area in more than 20 countries in the near east Africa and Asia. Farmers in these regions produce more than 50 per cent of the groundnut, 80 per cent of pearlmillet, 70 per cent of gram and 90 per cent of pigeonpea. The yields of these areas are extremely low as compared to those of humid and sub-humid regions. In some countries of Sub-Saharanian Africa and the near east food grain production per capita has declined significantly during the past decade. Although part of this decline can be attributed to high rates of population growth, periodic drought and unfavorable agricultural production and marketing facilities/ policies of the national governments. Much of it results from the steady and continuous degradation of agricultural lands from soil erosion and nutrient depletion and the subsequent loss of soil productivity. Many of these dryland areas are typified by a highly fragile natural resource base. Soils are often coarse textured, sandy and inhereditary low in fertility, organic matter and water holding capacity and easily susceptible to wind and water erosion. Runoff losses during rainfall events commonly exceed 50 per cent. Rainfall pattern are erratic and unpredictable and crops can suffer from moisture deficit and drought even during normal rainfall period. About 75 per cent of the crops produced in the near east region are grown under dryland or rainfed conditions. It is estimated that more than 70 per cent of the projected food and feed deficit by the year 2000 must come from increased yields on established crop lands since very little new arable land is available for agriculture development.

Table. Extent of non-irrigated lands to food production in selected countries.

Country	% cultivated land Non-irrigated	% total food from Non-irrigated lands
China	54	30
India	70	45
Indonesia	66	50
Pakistan	25	18

Key points of dryland areas

1. Arid and semi-arid regions comprise 40% of the world land area.
2. Some 700 million people inhibit this area.
3. Approx 60% of these drylands are in developing countries.
4. Low rainfed areas constitute from 75-100% of the land area in more than 20 countries in the near east, Africa and Asia.
5. These regions produce more than 50% of the groundnut, 80% of the pearlmillet, 90% of chickpea and 95% of the pigeonpea.
6. In India 40% of the total food production comes from drylands which supports 44% of the human population.

Why crop failures are common, yields are not static under rainfed farming because...

1. Inadequate and uneven distribution of rainfall.
2. Late on set and early cessation of rainfall.
3. Prolonged dry spells during the crop growth period.
4. Low moisture retention capacity of soils.
5. Low fertility of soils, low humidity, higher temperatures, higher wind velocity.

Distribution of rainfed areas in India

Three-fourth of our arable land (143 m ha) is rainfed. Food production is tied up with amount and distribution of rainfall; more than 90 per cent of the area under sorghum, groundnut and pulses is rainfed. In case of maize and chickpea, 82 to 85 per cent area is rainfed. Even 78 per cent of cotton area is rainfed. In case of rapeseed/mustard about 65.8 per cent of the area is rainfed. Interestingly, but not surprising 61.7, 44.0 and 35.0 per cent area under rice, barley and wheat, respectively, is rainfed. In terms of production, rainfed agriculture accounts for more than 40 per cent of the total food grains production, nearly 75 per cent of all oilseeds, 90 per cent of the pulses and 70 per cent of cotton.

Area under irrigated and rainfed conditions in India

Class	Area (Mha)	% total arable land	
Total arable	143.8	—	* includes flood prone area of
Dryland	34.5	24.0	15 M/ha
Rainfed	65.5*	45.5	
Irrigated	43.8	30.5	

Distribution of low rainfall areas in Haryana and India

In case of Haryana, the districts Sirsa, Hisar, Bhiwani and certain parts of Jind and Mahendergarh districts have an annual rainfall of less than 40 cm. While the districts like Kaithal, remaining parts of Jind, Rohtak, Mahendergarh, Rewari, Sonipat and some parts of Gurgaon, Faridabad, Panipat have annual rainfall of 40-60 cm. The rest parts of Haryana receive an annual rainfall of >60 cm. In India the low rainfall areas mostly lies in Western & Northern India & in Central Peninsular India.

1. Less than 20 cm rainfall : It includes Jaisalmer and western part of Bikaner district of Rajasthan. Certain areas of Gilgit and Ladakh Districts of Jammu and Kashmir also receive an annual rainfall of < 20 cm.
2. 20-40 cm rainfall : The central parts of Ladakh district of Jammu and Kashmir and Jamnagar & Kachchh districts of Gujrat receives annual rainfall of 20-40cm. The Bathinda& Ferozpur (Punjab), Sirsa, Hisar and Bhiwani (Haryana) and Barmer, Jodhpur, Nagaur, Churu, Sikar, Jhunjhunu, Shri Ganganagar and Hanumangarh (Rajasthan) districts all fall in this category of rainfall.
3. 40-60 cm rainfall : Jalor, Sirohi, Pali, Ajmer, Alwar and Jaipur districts of Rajasthan; Kaithal, Jind, Rohtak, Mahendergarh, Rewari, Sonipat, Gurgaon and Faridabad districts of Haryana; Amritsar, Faridkot, Jallandher, Ludhiana and Kapurthala districts of Punjab; Junagarh, Amreli, Rajkot, Mehsana and Banaskantha districts of Gujarat receives an annual rainfall of 40-60cm. Similarly Southern Western Ladakh District of Jammu and Kashmir also receives this rainfall. West Nimar District (Madhya Pradesh); Jalgaon, Aurangabad, Ahmadnagar and Sholapur Districts (Maharashtra); Bijapur, Raichur and Bellary Districts (Karnatka); Anantpur and Kurnool districts (Andhra Pradesh) and Periyar and Tirumelveli Districts (Tamil Nadu) also receives an annual rainfall of 40-60cm.

Table : State wise per cent unirrigated and irrigated area in India

S.No.	State	Irrigated area (%)	Unirrigated Area (%)
1.	Andhra Pradesh	43.9	56.1
2.	Arunachal Pradesh	24.9	75.1
3.	Assam	8.1	91.9
4.	Bihar	56.9	43.1
5.	Jharkhand	9.3	90.7
6.	Goa	17.5	82.5
7.	Gujrat	34.4	65.6
8.	Haryana	84.1	15.9

9.	Himachal Pradesh	19.2	80.8
10.	J&K	41.6	58.4
11.	Karnataka	29.2	70.8
12.	Kerala	18.7	81.3
13.	M.P.	43.2	56.8
14.	Chattishgarh	27.1	72.9
15.	Maharashtra	16.9	83.1
16.	Manipur	22.8	77.2
17.	Meghalaya	31.5	68.5
18.	Mizoram	17.4	82.6
19.	Nagaland	20.2	79.8
20.	Orrisa	32.2	67.8
21.	Punjab	94.9	5.1
22.	Rajasthan	38.7	61.3
23.	Sikkim	8.0	92.0
24.	Tamilnadu	56.4	43.6
25.	Tripura	21.8	78.2
26.	U.P.	78.6	21.4
27.	Uttrakhand	44.7	55.3
28.	West Bengal	59.2	40.8

Table : Extent of area under different crops in unirrigated or rainfed condition

Crop	Unirrigated area (%)
Total Pulses	83.8
Total Cereals	45.70
Total Oilseeds	72.90
Total Food grains	53.20

Table : Agro-ecological conditions in different rainfall zones in India

Annual rainfall (mm)	Soil type	Mean annual temp.	Length of growing season (days)	Dominant crops
<400	Sandy soil	0-10	30-90	Pearlmillet short duration Pulses
400-1000	Sandy, red,black	20-30	80-200	Pearlmillet, Sorghum, Maize, oilseed, pulses, cotton
1000-1800	Alluvial , laterite	20-30	200-300	Maize, Paddy, wheat, barley, Mustard, Sorghum,
>1800	Submountain	15-20	>300	Paddy, Plantation crop such as tobacco, tea

Table : Important dryland crops growing states in decreasing order

Crop	Rank-I	Rank-II	Rank-III	Rank-IV	Rank-V
Pearlmillet	Rajasthan	Maharastra	Gujrat	U.P	Haryana
Sorghum	Maharastra	Karnataka	M.P.	A.P.	Rajasthan
Chickpea	M.P.	U.P	Rajasthan	Maharastra	Haryana
Mustard	Rajasthan	U.P	M.P.	W.B.	Haryana
Cotton	Maharastra	Gujrat	Punjab	A.P.	Karnataka

Chapter 2

Rainfed Agriculture and Watershed Management: A Historic View

Rainfed Agriculture - A historic view

Historically, famines and scarcities have been recorded from the earliest times in India. They were mentioned in the epic Mahabharata, and there is Jain legend of a 12-year famine in the fourth century A.D. During the reign of the Moghul Emperor Shah Jehan in 1630-32, The official historian recorded in 'Badsha Namah' that soup Kitchens were opened, a lakh and half coins were given in charity and one eleventh of land revenue was remitted in the drought effected area.

In the remote past, these famines were tided over by saving grains and fodder from years of bumper harvest and by migration of people and cattle to greener pastures. With the rise of population pressure on arable land, the effects of scarcity and famine became progressively severe because the traditional methods of tiding over were no longer adequate. The successive famines in the closing decades of the 19th century closing decades of the 19th century led to the setting up of three Famine commissions.

The Commissions made several suggestions which formed the basis of the Famine Code promulgated in 1883 onwards and adopted by the then provincial governments. Apart from the relief works to be taken up during famine, the Commission recommended protective irrigation projects in dry and scarcity area and establishment of department of agriculture at the center and in the province. Later, the Royal commission on Agriculture (1928) reiterated in its report the problem of dryland agriculture and recommended a close attention by departments of agriculture to these problems.

The first systematic approach was made in 1923 by setting up small plots at Manjri, near Poona. This was followed by a more comprehensive research scheme on dry farming sanctioned by the Imperial (then Indian) Council of Agricultural

Research during 1933-35. Under the scheme, research was conducted in low rainfall areas at Rohtak, Solapur, Bijapur, Raichur and Agra. Based on the results of this scheme, a series of dry farming practices were evolved which were known as ' the Bombay Dry farming ', ' the Madras Dry farming ' and ' the Hyderabad Dry farming ' practices. These practices emphasized soil and water conservation as basic to dryland production programs. The Scheme was terminated in 1943, since it was felt that the immediate objectives had been achieved.

After independence, nine soil conservation research centers were set up in different parts of the country in 1954. Large-scale action programs in soil and water conservation were taken up which benefited largely the dry farming areas. The rainfed area, which is 75% of the cultivated, contributes 42% to the nation's food. Grain legumes, oilseeds and cotton are almost exclusively grown in drylands. Rainfed rice is also a dominant crop of humid region. Recognizing this, an All India Coordinated Research Project for Dryland Agriculture was initiated in 1970. Financial support came from the Government of Canada. The main goals were to reduce the growing socio-economic gap between the drylands and less water endowed regions and to establish efficient patterns of land utilization and moisture conservation.

Status of rainfed agriculture in India

"There was a time when it was said that the Indian budget was a gamble on the monsoons. That is not the case anymore, with industrial production soaring and agriculture on the margins. But what is clear is that without addressing the problems of the majority of farmers in unirrigated land (where medium of irrigation is none except rain), we are not going to achieve food security." Rain-dependent areas can be broadly split into two: 'dry lands', which receive less than 750 mm (75cm) of rain a year; and rainfed areas, which receive more than 750 mm. (75cm). Comprising arid and semi-arid ecosystems, dry lands stretch from Gujarat in the west till Eastern Madhya Pradesh; and from Rajasthan till the southern tip of India. India ranks first among the rainfed agricultural countries of the world in terms of both extent and value of produce. Due to population pressure on agricultural lands, the *poverty is concentrated in rainfed regions.* The climate in India's rainfed regions is characterized by complex climatic deficiencies, manifested as water scarcity for rainfed crop production. The climate is largely semi-arid and dry sub-humid with a short (occasionally intense) wet season followed by long dry season. Rainfall is highly unreliable, both in time and space, with strong risks of dry spells at critical growth stages even during good rainfall years.

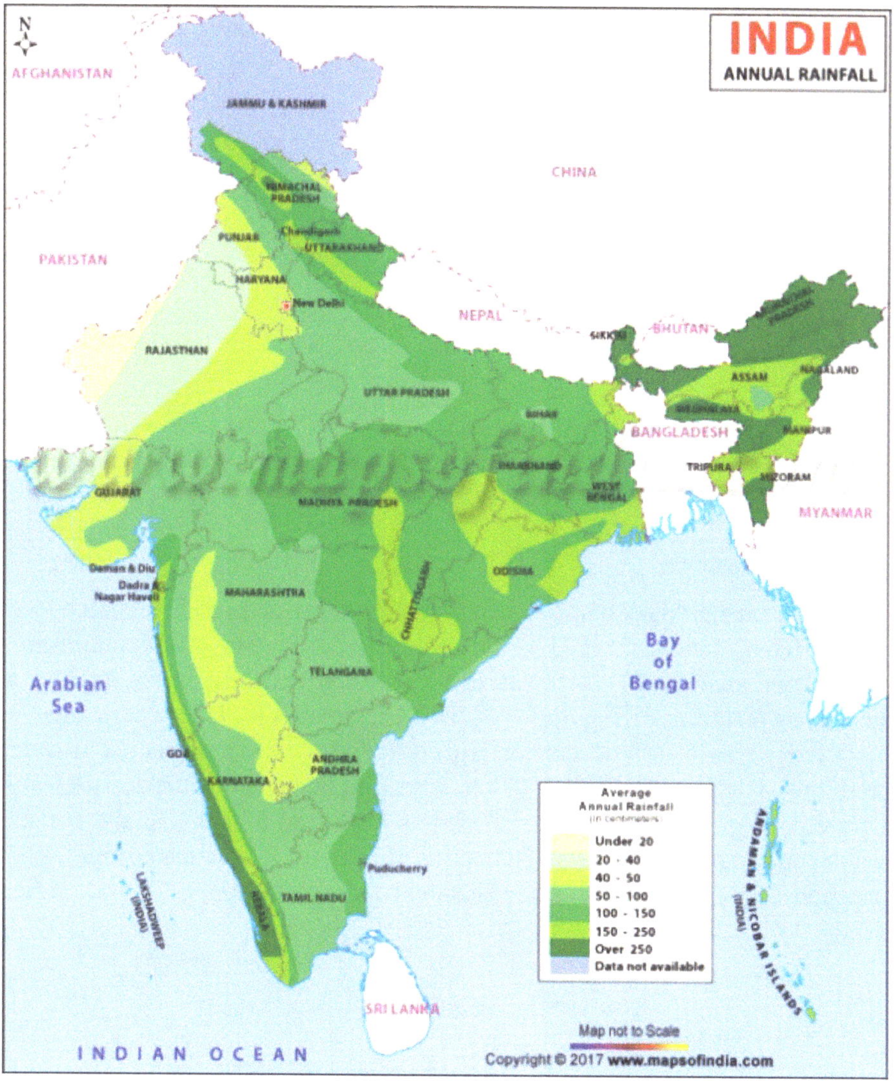

Figure 1

Above is the average annual rainfall map of India (Figure 1).

There has been deficient rainfall over past few years and that has affected availability of water for agriculture. Figure 2 shows this data.

Rainfall in the country (as compared to the average rainfall since 1951) has been volatile over the past 20 years. The lowest level of rainfall was in 2009-10, at 22% below normal, whereas the highest level of rainfall was in the years 2013-14, 2007-08 and 1998-99, at 6% above normal. This deviation in rainfall may not be uniform across the country. For example, in a year of overall excess rainfall, there may be regions which received deficient rainfall.

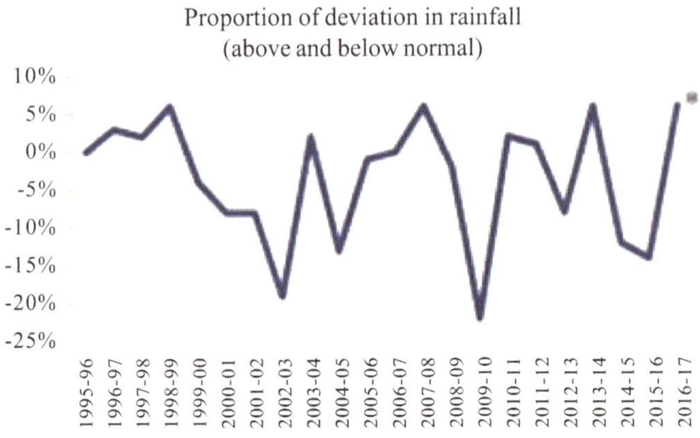

*estimate of Indian Meteorological Department for 2016 season.
Note: Above 0% - surplus in rainfall; below 0% - deficiency in rainfall

Figure 2

Present scenarios

It is estimated that "more than half of agricultural sown land, i.e. 53% is dependent on rain-fed irrigation. 53% (73.83 million hectares) of the sown agricultural area (total sown area for 2012-13 was 139.93 million hectares) in the country is dependent on rain-fed irrigation as of 2012-13, which is the latest data available. Rain-fed irrigation is dependent on rain water and is not supplemented by water from any other source. Rest 47% (66.1 million hectares) of the sown area is irrigated by ground water sources such as wells and tube-wells, and surface water sources such as canals and tanks, as of 2012-13 [Net irrigated area]. This area constituted 18% (20.9 million hectares) of the sown area in 1950-51."4 See Figure 3.

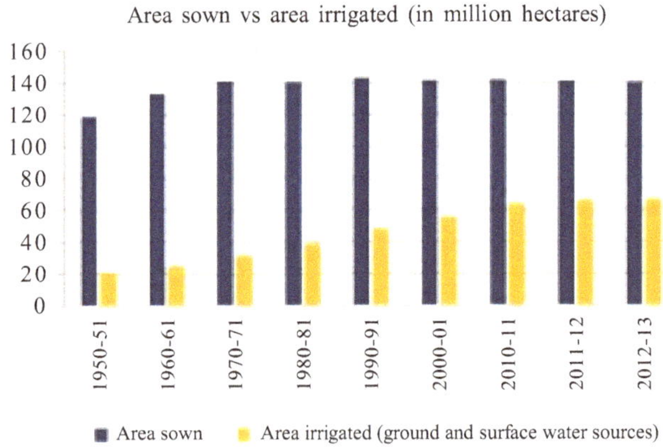

Figure 3

It is mostly the North Eastern states that depend on rainfall while Northern states depend on ground and surface water. "The northern states of Punjab (99.5%), Haryana (90%), and Uttar Pradesh (76%) have the highest proportion of sown area which is irrigated by groundwater and surface water sources. In contrast, less than 25% the sown area in the north-eastern states of Manipur, Meghalaya, Mizoram, and Nagaland is irrigated by ground and surface water sources. This implies that these states rely more on rainfed irrigation. The central states such as Maharashtra (17%) and Chhattisgarh (26%), and southern states such as Kerala (22%) and Karnataka (27%) also have lower proportions of sown area under irrigation."

Conclusions on status of water availability for agriculture in india

In the first part of the article on agriculture, we explored one of the major sources of irrigation in India, i.e., dams and canals. In this part we have seen the status of the next two major sources of irrigation, i.e., groundwater and rain. As we saw above 53% of the sown agricultural area in the country is dependent on rainfed irrigation as of 2012-13 and the rest 47% is dependent on groundwater. In figure 5 below of the "sources of irrigation." 6

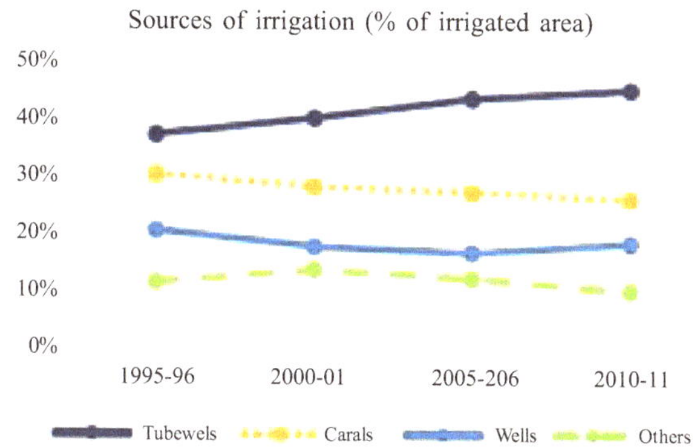

Figure 4

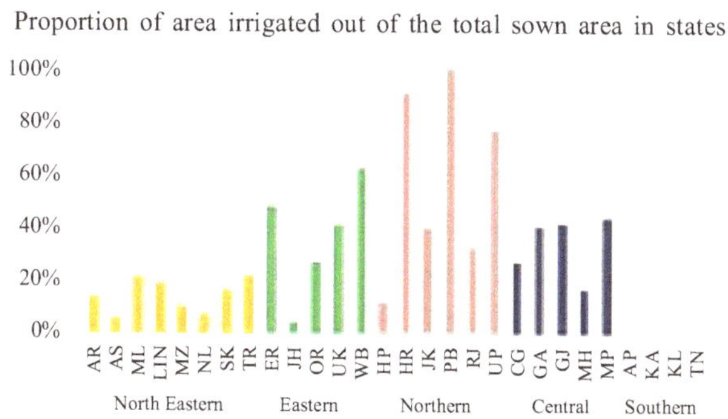

Figure 5

For complete data on land use and irrigation pattern see Table 1 below:

Table 1: Pattern Of Land Use And Irrigation (Area in million hectares)

Year	Net Sown Area	Gross Sown Area	Net Irrigated Area	Gross Irrigated Area
1	2	3	4	5
1990-91	143.00	185.74	48.02	63.20
1991-92	141.63	182.24	49.87	65.68
1992-93	142.72	185.70	50.29	66.76
1993-94	142.34	186.58	51.34	68.26
1994-95	142.96	188.05	53.00	70.65
1995-96	142.20	187.47	53.40	71.35
1996-97	142.93	189.50	55.11	76.03
1997-98	141.95	189.99	55.21	75.67
1998-99	142.75	191.65	57.44	78.67
1999-00	141.06	188.40	57.53	79.22
2000-01	141.34	185.34	55.20	76.19
2001-02	140.73	188.01	56.94	78.37
2002-03	131.94	173.89	53.90	73.06
2003-04	140.71	189.66	57.06	78.04
2004-05	140.64	191.10	59.23	81.08
2005-06	141.16	192.74	60.84	84.28
2006-07	139.82	192.38	62.74	86.75
2007-08	141.02	195.22	63.19	88.06
2008-09	141.90	195.33	63.64	88.90
2009-10	139.17	189.00	61.94	85.08
2010-11	141.56	197.56	63.66	88.93
2011-12	140.98	195.69	65.70	91.78
2012-13	139.94	194.14	66.27	92.25
2013-14	141.43	200.86	68.10	95.77

Watershed management – a historic view

Large-scale removal of forest lands by humans in the nineteenth and early part of the twentieth century's created significant changes in the hydrologic function of watersheds. Downstream flooding occurred more frequently, with subsequent increases in loss of life and damage to infrastructure. Accelerated erosion, produced by changes in the biotic and hydrologic components of natural drainages (watersheds), created unprecedented large-scale siltation of developed lowlands. At the time, the general consensus was that the removal of forest was causing these undesirable impacts. However, the mechanisms for reversing the process through sound scientific management had not been developed. During the second quarter of the twentieth century, the discipline of forest hydrology evolved from the need for scientific management of the soil and water resources of headwater catchments in order to minimize the flooding and siltation of productive lands and infrastructure in the valleys and plains inhabited by humans. As the importance of rangelands and cultivated lands in the hydrologic cycle and the erosion–sedimentation processes of catchments became known, forest hydrology gave way to the more comprehensive, present day watershed management. Over time and in response to changing needs, the scope of watershed management has broadened from the initial concept of technical management of the water resource to an integrated discipline that applies biological, technical, social and economic principles to maintain the productivity of headwater and lowland areas through the scientific management of soil, plant and water resources.

Impacts of watershed management technologies

Watershed management technologies have proven to be effective for mitigating erosion on sloping land, stabilizing landscapes, providing clean water, and stabilizing – and in some instances improving – agrarian production systems on the small to medium scale. With modification, these existing technologies can be used successfully in most terrestrial environments inhabited by humans. The degree of success of watershed management interventions is primarily a matter of the will of the people and the scale of the activities. Regarding the *upstream* effects, examples exist throughout the world where upland resource conservation activities have been successful on the micro and macro scales; e.g. micro- to mesoscale activities in Honduras, the Philippines, China, Thailand, Burundi, Nepal, Pakistan, Sri Lanka, India, Bolivia, Peru and other countries, and the macro-project in Santa Catarina, Brazil. Regarding the effects *downstream*, the impact of upland watershed management activities on downstream water quantity, quality and siltation remains a controversial issue, partly because of economies of scale, and partly because of difficulties in predicting with reasonable accuracy the results of these activities. Until the magnitudes of natural and human-induced erosion and subsequent sedimentation can be quantified with reliability in a watershed, the controversy will remain regarding upstream effects on downstream infrastructures. The same applies to the quantifiable affects of land use on the hydrologic cycle and water supply and quality. In the meantime, downstream infrastructures such as hydroelectric and/or irrigation dams are being constructed for hundreds of millions of dollars. However, in the past, when watershed management activities were to be carried out to mitigate downstream siltation of these structures, at best a few million dollars were provided to treat all of the contributing upland areas. In many catchments, the upland areas are in degraded condition before the dam is constructed, so implementing small-scale watershed management interventions is like putting a band.

Watershed management: Water resources for the future

Watershed management in its truest form is the conservation management of the soil, plant and water resources of a catchment to benefit humanity. It involves managing the land and human resources of the drainage in a manner that sustains adequate levels of water, soil, food and fibre production. This form of management requires a participatory integrated approach that includes the various physical, vegetative and human components of areas that range from a few hectares to large river basins. The watershed part of watershed management implies management of these resources, to the extent possible, within a defined physiographic boundary. From a conceptual perspective, when the boundaries of a management system are defined it is easier to identify and monitor the components (e.g. inputs, storage and outflows) of that system – e.g. the hydrologic cycle. However, from a land management perspective, these physical boundaries are considered to be simply topographic demarcations within political and administrative boundaries that usually overlay a series of watersheds.

The theoretical concept of participatory integrated management of natural resources is difficult to apply. The myriad uses, ownerships, political and social constraints and biophysical systems in large watersheds limit application of the idealistic integrated approach. In practice, large catchments are usually managed according to economic, social and political considerations. Management of the natural resources in headwater watersheds has the greatest potential for application of the participatory integrated concept. Agricultural, forest and rangelands often represent a potentially significant production resource for local inhabitants. However, the natural physical and biological constraints of uplands often limit productivity compared with lower elevations where major production and population centres are located.

Watershed management

A *watershed* is the area that drains to a common outlet. It is the basic building block for land and water planning. *Degradation of watersheds* in recent decades has brought the long-term reduction of the quantity and quality of land and water resources, as shown in the cases of Lesotho and Morocco. Changes in watersheds have resulted from a range of natural and anthropogenic factors, including natural soil erosion, changes in farming systems, over abstraction of water, overgrazing, deforestation, and pollution. The combination of environmental costs and socioeconomic impacts has prompted investment in watershed management in many developing countries.

Watershed management is the integrated use of land, vegetation and water in a geographically discrete drainage area for the benefit of its residents, with the objective of protecting or conserving the hydrologic services that the watershed provides and of reducing or avoiding negative downstream or groundwater impacts. The key characteristics of a watershed that drive management approaches are the integration of land and water resources, the causal link between upstream land and water use and downstream impacts and externalities, the typical nexus in upland areas of developing countries between resource depletion and poverty, and the multiplicity of stakeholders. Watershed management approaches need to be adapted to the local situation and to changes in natural resource use and climate.

Lessons learned from watershed management projects in the 1970s and 1980s

The first generation of watershed management projects in developing countries in the 1970s and 1980s applied a soil and water planning approach to watersheds that emphasized engineering works aimed at specific on-site and downstream physical outcomes. Less attention was paid to the needs of upstream populations or to their ownership of program actions. As a result, investment costs were high and not always well justified, and the assets and benefits created often had a limited life. By the end of the 1980s, the comparative failure of this "engineering-led" approach was clear, and a major rethinking of watershed management approaches was undertaken by national and international agencies.

The integrated and participatory watershed management approach of the 1990s

The 1990s represented a new departure for watershed management programs supported by the international community in developing countries. Although engineering solutions were not excluded where appropriate, the emphasis was placed more on farming systems and on participatory and demand-driven approaches implemented at the decentralized level. Impetus was given to this new departure by the renewed emphasis on rural poverty reduction in development programs. The move away from planned investments toward farming systems and participatory approaches was designed to seek synergies and to limit the need for tradeoffs. However, it posed two considerable challenges. First, it was not clear under what circumstances the new watershed management approach could achieve both conservation objectives *and* income increases. Second, it remained to be demonstrated whether investments made upstream under a demand driven watershed management program could have a positive effect on downstream conditions by improving hydrological services or reducing negative externalities.

The evolution of national policies in watershed management

National policies on watershed management have tended to develop in a pragmatic and iterative fashion, with early setbacks over engineering-dominated approaches being succeeded by tests of community-based approaches and by technology packages targeting sustainable changes in land and water use practices that brought profit to stakeholders. In several countries, including Brazil, China, India, and Turkey, success in testing community-based approaches has led to adoption of broader policies for community-based watershed management. In other countries, including Morocco and Indonesia, doubts about program performance and cost have delayed the adoption of national policies.

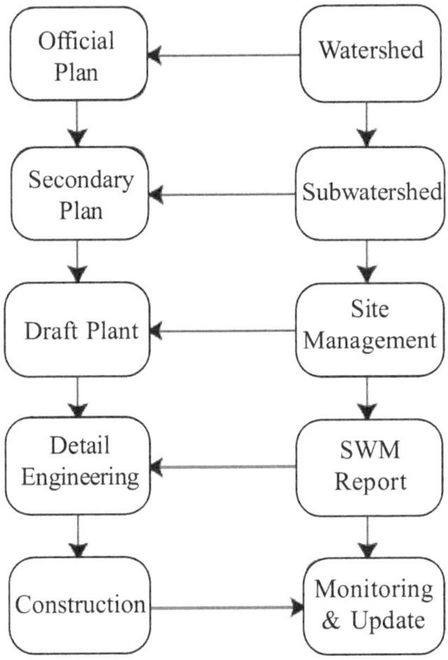

Participatory processes in the planning and implementation of watershed management activities

Global experience has shown that there is no universal model for participatory planning and implementation of watershed management activities. There is a process that would, in most cases, have similar steps. However, this process – which should include all levels and steps of the participatory process, e.g. planning, design and implementation with all concerned stakeholders – has not been well defined. Bits and pieces of the process have been identified by various projects. The complete participatory process for watershed management needs to be mapped out in a logical manner, tested and refined. Experience has shown that empowerment of the main stakeholders in watershed management projects/ programmes to plan and implement appropriate activities is essential if the project/

programme is to have any chance of sustainability. For example, regardless of good intentions, it is not enough for a project to form a community conservation committee at the grassroots level – in isolation from local governments – plan and start interventions, provide technical, financial and other required inputs to the end of the project and then expect the government to make the project sustainable by providing the required inputs into the future. This is a recipe for failure. Participatory research methods such as participatory rural appraisal, which have been developed and employed on a wide scale in watershed management projects, have sometimes been a good instrument for initiating the participatory process. However, owing in part to the inherent nature of rapid data collection, subjective questions and answers and limitations on statistical analyses and the subsequent extrapolation of findings, the data generated by these rapid survey methods have limitations for use as baseline data for future assessment of project success. In addition, these participatory appraisal methods are only one part of the participatory process.

Participatory approaches and institutional considerations

The pendulum is swinging in support of empowering people with regards to the conservation of natural resources. There are several reasons for this, one being that past endeavours by governments to solve natural resources degradation problems on their own have for the most part been unsuccessful in terms of sustainability. Second, most national governments do not have the human or financial resources for the countrywide mitigation of natural resource degradation. Throughout the world there are examples of successful, sustainable resource conservation being carried out by local communities that have been empowered to manage their land-based resources. Change is also occurring, albeit slowly, in governments. New policies are being implemented that permit and encourage people's management of their natural resources; e.g. land tenure, user rights, water rights, crop tenure, formal recognition of community groups and committees, privatization of communal lands, rights to the income generated from these conservation activities, *etc*. The participatory process requires an active, well-trained field-level extension service in sufficient numbers to carry out watershed management activities on a large scale. The extension component is usually a weak link in the development process.

Chapter 3

Problems and Prospects of Rainfed Agriculture in India

Problems Associated with Rainfed Farming Agriculture

It is a dream desire to eliminate drought from this region. There are many problems that are associated with rainfed farming systems. The problems may be grouped into four major factors: climatic, edaphic, technological and socio-economic.

Climatic

1. Scarcity of rain-water which is low and unpredictable with respect to intensity and distribution;
2. High *ET* (nearly 3650 mm annually compared to total annual precipitation ranging from 350 to 1400 mm);
3. Extreme thermic values (upto 49^{0}C with the mean annual temperature exceeding 18^{0}C at some places);
4. High solar incidence (450 to 500 Cal cm^2 day^{-1});
5. High wind velocity with desiccating winds causing a high rate of *ET* and wind erosion;
6. Low relative humidity;
7. Extensive climatic hazards such as weather aberrations, drought, flood, frost, gale, cyclones and burning winds due to dry, deserted and denuded situations.

Edaphic

1. Poor and marginal lands with soils low in fertility and productivity;
2. Uneven topography with high erodability;
3. Difficulty in workability particularly in vertisols;

4. Shallow or very deep in depth with extreme permeability;
5. Low moisture storage and release capacity particularly in Alfisols;
6. Presence of dissolved injurious salts in ground water;
7. Problem soils with respect to soil reaction (pH) and high concentration of soluble salts in the surface soils;
8. Water logging in level lands; flooding and breaking small field bunds resulting in poor conservation of soil and water;
9. Movement of sand and soil;
10. High surface crusting that leads to poor crop stand and high rate of evaporation and mechanical injury to roots.

Technological

1. Limited choice of crop varieties matching the short moist period suitable for cropping as about 60 per cent of the area has two to four-and-a-half wet months and about 19 per cent of the area has less than two wet months per year.
2. Difficulty in designing a suitable cropping system, crop mixture and crop geometry considering soil type and other flexible components such as the moisture availability index, and the duration of the monsoon;
3. Difficulty in evolving static and adaptable technologies fitting into the prevailing as well as anticipated rainfall conditions on a regional basis. The use of high input based technology is risky;
4. Difficulty in seed production and multiplication specially in an on-far basis;
5. Farming is mainly human and animal power-based which is labour intensive and land saving but costly and time consuming resulting in difficulty in covering the entire arable area under the brief period of favorable weather during the season;
6. Limited use of fertilizers, poor response of biofertilisers and poor availability and the scope to improve it is very restricted;
7. Poor land capacity and the scope to improve it is very restricted;
8. Limited scope of use of residual moisture and nutrients;
9. Difficulty in conserving moisture for proper and timely utilization by crops;
10. Difficulty in reclaiming problems soils;
11. Lacking light and speedy implements with cheap sources of energy;

12. Profuse weed, pest, pathogen and parasite infestation;
13. Unpredicted heavy showers resulting in the accumulation of more than optimum moisture in the soil, impairing field operations including sowing interculture and harvesting in time;
14. Limited scope of land improvement;
15. Water harvesting and recycling is the high investment proposition at the initial level;
16. Water damage and loss of run off (up to 50 per cent of incident rainfall) causing loss of soil (greater than 10t/ha) and formation of gullies resulting in degradation and unsuitability for arable cropping.

Socio-economic

1. Peculiar ecological and socio-economic settings;
2. Frequent failure of crops and unstable production rendering farmers poorer;
3. Low cropping intensity; low farm income, malnutrition and poor quality of drinking water;
4. Small size of farm holdings and high population pressure on land (1.7 persons/ha);
5. Poor quality of produce fetching lower price;
6. Lack of marketing facilities and market incentives;
7. Low level of literacy and poor resource base of the farmers through they are economic minded and rational;
8. Unemployment for most of the period of the year;
9. Farmers are to depend on the favours of the monsoon or on financing institutions which are reluctant to provide assistance as there is a more risk in the recovery of the released amount from farming;
10. Farmers of these regions are deprived of innovations, initiative, inspiration, aspiration and appropriate incentives and appreciations and they wait, watch and worry over their future.

Crop wise production constraints

(A) Bajra

1. Use of lower seed rates and untreated seeds

   Farmers must be educated by laying down demonstration on their fields and analyzing the additional costs and additional grains from recommended levels of seed rates as well as treated seeds.

2. Non-adoption of chemical fertilizers including the micronutrients

   By laying down demonstration on farmers' fields, they must be educated about the use of adequate fertilizers under the rainfed condition.

3. Thinning and gap filling

   The technical feasibility and economic benefits of these practices must be demonstrated to farmers by conducting demonstrations trials on their fields.

4. Non-adoption of sowing by Ridger seeder

   Tractor drawn ridger seeder is very heavy and costly and there is no equally efficient and good ridger seeder drawn by bullocks.

5. Weed control

   To ease the problem of labour shortage, its efficiency should be increased through the use of more efficient implements, tools simple machines; such tools must be developed/popularized on priority basis.

6. Plant protection

   Suitable varieties resistant to major insects, pests, and diseases must be evolved. Less costly and adequately effective insecticides/pesticides weedicides and other plant protection chemicals must be evolved for popularization among the farmers.

7. Lack of varieties for dry farming conditions

   There is strong need for developing suitable HYV highly tolerant to moisture stress conditions.

8. Soil moisture

   Under the dryland farming situations most of the crops suffer from moisture stress. Appropriate technology for conserving the soil moisture as well as other low cost water harvesting technologies must be evolved and popularized.

(B) Clusterbean

1. Use of improved varieties

   Farmers do not have knowledge about improved varieties that have substantially higher yield potential than the local ones.

2. Use of rhizobium culture

   Its utility is shown on paper but most of the farmers are not convinced about it. It needs to be demonstrated on farmer's field to convince them.

3. No/negligible use of fertilizers

   The significance of the balance use of fertilizers on yield and profitability of guar needs to be demonstrated by conducting trials on farmer's fields.

4. Inter culture and plant protection operations are not common

   The effect of inters culture and plant protection operations on yield and profit needs to be demonstrated by the extension agencies to convince the farmers.

(C) Wheat

1. Lower doses of fertilizer application

   By conducting demonstrations/trials on farmer's field the economic benefits particularly the additional grins to the farmers due to use of full recommended doses of fertilizers needs to be demonstrated to convince them.

2. Irrigation at right time

   Farmers are well convinced about the timely irrigation in wheat their constraints pertaining to non-availability of adequate electricity/canal water be solved by the Govt. to the maximum possible extent.

3. Effective weed control measured

   Farmers are not fully convinced about the magnitudes or increase in yields/ profit by applying adequately weed control measures/interculture operations in wheat.

4. Non-adoption of various plant protection measures recommended by the technical scientists

   Most of chemicals recommended for plant protection measured are very costly. Research efforts are needed to find out cheaper and equally effective chemicals for this purpose. Also, most of the chemicals are very costly. Farmers need to be convinced about recommendations concerning the plant protection measures.

5. Non-adoption of weed control measured by majority of the farmers

   Farmers need to be educated to be educated to convince them about substantial economic benefits of controlling weeds in their fields. Demonstration on farmer's field must be laid out and its economics be worked out to convince them.

(D) Gram

1. Adoption of improved varieties

   Relatively superior varieties to common ones are available and well planned yield maximization demonstration of these recommended varieties needs to be laid out on farmer's field.

2. Use of rhizobium culture

   To convince the farmers about the importance of inoculation of seed with rhizobium culture in increasing the yield.

3. Seed treatment for protection against termites

   The relevant chemicals must be made easily available to the farmers and they must be educated for treatment of seeds before sowing.

4. Uneven germination and poor plant population due to moisture stress at planting time

   The available technology for the conservation of soil moisture must be popularized among the farmers by the extension agencies through laying out field demonstration on the farmer's fields.

5. Very low use of chemical fertilizers

   Farmers have apprehension that use of fertilizers under rainfed conditions adversely affects the yield. Also many farmers were not aware of the importance of the phosphatic fertilizers.

(E) Barley

1. Non-use of recommended varieties

   In farmer's opinion whatever improved varieties are available are not significantly superior to the locally common varieties to convince the farmers about the superiority of recommended varieties demonstrations be laid out on farmer's field and its comparative economic be worked out.

2. Application of very low dose of chemical fertilizers

   Farmers are not fully convinced about the application of true recommended doses of fertilizers under their condition not having adequate irrigation water.

3. Inter culture and weed control measures

   Farmers do not adhere to this practice because they are not quite convinced about its economic advantages. They need to be convinced through laying down demonstrations on their fields and working out its comparative economics.

4. Seed treatment against seed borne diseases and termites

   Farmers need to be educated about its significant advantages through laying down trials/demonstrations on farmer's field.

(F) Raya (rapeseed and mustard)

1. Thinning of crop to make optimum spacing for plants

   Farmers need to be convinced about its economic advantages by conducting trials on their fields.

2. Application of very low doses of fertilizers

   Field demonstrations must be conducted on typical farm conditions to demonstrate the economic benefits of application of the recommended dose of fertilizers as most of the farmers are not quite convinced about these recommendations.

3. Inter culture and weeding

   Farmers are not very convinced about its substantial economic gains. It needs to be demonstrated.

4. Insect, pests and disease control measures

   Most of the farmers lack technical know-how about various recommended plant protection measures. Demonstrations must be conducted on farmer's field and its comparative economic be worked out to educate them.

Constraints limiting crop production in dryland areas

Dry farming areas are characterized by very low and highly variable and uncertain yields. Crop failures are quiet common. These are mainly due to the following causes:

1. Inadequate and uneven distribution of rainfall

   In general, the rainfall is low and highly variable which results in uncertain crop yields. Besides its uncertainty, the distribution of rainfall during the crop period is uneven, receiving high amount of rain when it is not needed and lack of it when crop needs it.

2. Late onset and early cessation of rains

   When the monsoon sets in late, the sowings of the crops are delayed resulting in poor yields. At times, the rains may cease very early in the season exposing the crop to drought during flowering and maturity stages which reduces the crop yields considerably.

3. Prolonged dry spells during the crop period

   Long breaks in the rainy season are an important feature of Indian monsoon. These intervening dry spells when prolonged during crop period reduces crop growth and yield and when unduly prolonged crops fail.

4. Low moisture retention capacity

   The crops raised on red soils, and coarse textured soils suffer due to lack of moisture whenever prolonged dry spells occur due to their low moisture holding capacity. Loss of rain water occurs as runoff due to undulating and sloppy soils.

5. Low fertility of soils

   Dry lands are not only thirsty, but also hungry too. Soil fertility has to be increased, but there is limited scope for extensive use of chemical fertilisers due to lack of adequate soil moisture.

In India, the problem of crop production in dry lands are more numerous and varied as it is practiced under a wide range of climate and soil conditions. These problems can be broadly grouped into three: Vagaries of monsoon, soil constraints and socio- economic.

Vagaries of monsoon

Based on average annual rainfall, the country can be divided into three zones:

1. Low (< 750 mm)
2. Medium (750-1150 mm)
3. High (>1150 mm) rainfall zones.

Dryland area is nearly equally distributed among the three. Areas with <1150 mm (arid and semiarid) are the problem areas for crop production. Main characteristics of rainfall influencing crop production are its variability, intensity and distribution, late onset of monsoon, early withdrawal of monsoon and prolonged dry spells during the crop period.

1. Variable rainfall : Annual rainfall varies greatly from year to year. Generally, higher the rainfall less is the coefficient of variation. Crop failures due to uncertain rains are more frequent in region with lesser rainfall.
2. Intensity and distribution **:** In general > 50% of total rainfall is usually received in 3 -5 rainy days. Such intensive rainfall results in substantial loss of water due to surface runoff. This process also accelerates soil erosion. Distribution of rainfall during the crop growing season is more important than total rainfall in dry land agriculture.
3. Late onset of monsoon : If the onset of monsoon is delayed crops/varieties recommended to the region cannot be sown in time. Delayed sowing leads to uneconomical crop yields.
4. Early withdrawal of monsoon : This situation is equally or more dangerous than late onset of monsoon. Rainy season crops will be subjected to terminal soil moisture stress leading to poor yields. Similarly, a post-rainy season crop fails due to inadequate available soil moisture, especially during reproductive and maturity phases.
5. Prolonged dry spells : Breaks of monsoon for 7-10 days may not be a serious concern. Breaks of > 15 days duration, especially, at critical stages for soil moisture stress leads to reduction in yield. Drought due to break in monsoon may adversely affect the crops in shallow soils than in deep soils.

Soil constraints

Alluvial soils occupy the largest area in dry land agriculture. Problems of crop productions are not as acute in these soils as they are in black and red soils. Major problems are encountered in Vertisols, Alfisols and related soils. Black (Vertisols), red (Alfisols) and associated soils are mostly distributed in central and south India. The coastal areas have Alfisols, laterite and lateritic soils.

Alfisols

These are commonly referred as red soils. Problems related to crop production are:

1. Poor crop stand due to crusting and rapid drying of surface soil.
2. Poor crop growth due to unreliable soil moisture supply, low moisture storage capacity due to shallow depth and drought spells during crop season.
3. Low soil fertility due to low organic matter, poor nutrient status particularly with respect to N, P, S and compact subsoil layer (argillic horizon).
4. Land degradation from soil erosion and crusting.

Vertisols

These soils are commonly called as Black soils and characterised by high clay content (30-70%). Important constraints for crop production are:

1. Physical constraints such as narrow range of soil water content for tillage, tendency to become waterlogged and poor trafficability.
2. Low soil fertility due to low N and available P.
3. Land degradation from soil erosion and salt accumulation, especially in low-lying areas.

Inseptisols and entisols

These are commonly termed as alluvial soils.

1. They have low water holding capacity.
2. Management for these soils for crop production is relatively easy compared to red and black soils.
3. Soil erosion is, however, a problem leading to land degradation.

Prospects of rainfed agriculture in India

The Indian economy is mainly dependent on agriculture, which contributes 21percent of country's capital GDP and 60 percent of employment potential. India made rapid strides in food production during last three decades culminating in self-sufficiency and surplus production. However, feeding the ever-increasing population through the next millennium remains an uphill task. The country will have to feed about 1.3 billion people by the year 2020 requiring 5-6 mt of additional food grains every year. Besides, the problem of poverty and malnutrition has their own implication to national food security. Over 70 percent of Indian population, which is predominantly rural, do not have proper access to food and nonfood commodities due to poor employment and infrastructure facilities. This reminds all those concerned with the country's food security of the daunting task ahead in order to ensure access to food to the growing population. Rainfed agriculture occupies 67 percent of net sown area, contributing 44 percent of food grain production and supporting 40 percent of the population. Even after realization of full irrigation potential of the country, 50 percent of net sown area will continue as rainfed. At present 95 percent of area under coarse cereals, 91 percent under pulses. 80 percent under oilseeds, 65 percent under cotton and 53 percent under rice is rainfed. Livestock forms an integral part of rainfed ecosystem and two out of every three animals are thriving in these regions. These areas are spread-out throughout the length and breadth of the country with semi-arid to sub-humid environments, shallow textured light soils to deep textured black and alluvial soils with varied effective crop growing periods from 90 to 180 days. Keeping the importance of rainfed agriculture in mind we have to give more emplhasis to these areas through technological development and extensive. Second Green Revolution must come in these grey areas.

Scenario of food demand and resources

The food grain requirement of the country is around 250 mt, out of which food demand could be about 105 mt of rice, 85 mt of wheat, 35 mt of coarse grains and 22 mt of pulses, 10 mt of oilseeds and 120 mt of milk and 110 mt of vegetables and for fruits 70 mt. The food grain requirements have been projected for 2025 at 308 mt with low population growth of 1286 million. But at higher population growth scenario (1333 million), the projected food grain production is 320 mt by 2025. More than the calories, ensuring protein security will become an important issue in view of the predominantly vegetarian habits of the populace and the dwindling availability of vegetable (pulses) proteins whose current supply is about 25 g head^{-1} day^{-1} against the minimum dietary need of about 70 g. The agriculture production increased from 50 mt to over 200 mt, between 1950-2000, thanks to green revolution. This, however, had its own costs in terms of degradation of land and water resources, loss of plant biodiversity, shift of agricultural land to non-agricultural uses, polluted environment, widening gap between the rich and the poor. Thus, physical access to food was the most important food security challenge in the past but economic and access to food has now become the most important cause of hunger and ecological access to food might become the most important concern in the next millennium owing to the damage now being done to land, water, flora, fauna and atmosphere.

Shrinking of natural resources

The per capita availability of agricultural land in India was 0.46 hectares in 1951 which decreased to 0.15 hectares in 2000 as against the global average of 0.6 ha. Number of persons per hectare of net-cropped area was 3 in 1951, 6.5 in 2000 and is estimated at 8 persons in 2025. This situation of rapidly declining land to man ratio is likely to worsen further owing to competitive demand for food, fibre, fuel, fodder, timber and developmental activities such as urbanization and industrialization, special economic zones,mining, road construction and reservoirs *etc.*

Chapter 4

Climatic and Soil Conditions Prevalent in Rainfed Areas

Climatic conditions prevalent in rainfed areas

Weather, which is part of climate, plays an important role in crop planning in dry farming area. Out of the several elements of weather, rainfall has key position in success of dry farming. In dryland areas, South West Monsoon brings the bulk of rainfall. The South West Monsoon is followed by North East Monsoon which supplements to South West Monsoon are the main source of rainfall. There are four types of rainfall characterized by the nature in different parts of India. Generally, the rainfall is scanty, erratic and ill distributed.

1. Rainfall

The monsoon period is not one of continuous rains in any part of the country. Meghalaya that is uncommon in the behavior of monsoon in the country. There are at least four variations in rainfall.

1. The commencement of rains may be quite early or considerable delayed according to the dates of monsoon for central Maharashtra zone is 10th June but is started as early as 28th May 1925 and its onset was delayed upto 26th June 1954 according with the date of onset of monsoon in Rajasthan is between June 21 to 30 but is started as late as up to end of July 1987 to 1992.
2. There may be prolonged breaks in rainfall. These breaks in monsoon rains can be different duration. The breaks of 5-7 days may not be all serious concern but breaks of longer duration of 2-3 weeks created the situation of plant water stress leading to reduction in agricultural production. Another important point is the physiological properties of soil and available water holding capacity in the relation to the breaks and drought, some of the soils e.g. deep black soils have capacity to store more than 300 mm of water per

meter depth where as the desert soil can ensure only 100 mm. The impact of drought is felt more quickly in the soil having less water holding capacity but in sols with high storage capacity 15 days dry break may not affect the crop where as the soils of less storage capacity show wilting symptoms in and dry break such breaks cannot be predicted for a given season in advance. The drought caused due to this break is therefore of different magnitude and severity.

3. The rains may terminate considerably earlier or persists long than usual. The problem of withdrawal of monsoon is similar to onset of south-west monsoon. The date of withdrawal of south-west monsoon in Rajasthan is between September 19–20 its withdrawal was as early as 15th August in 1987 and 1989. The drought of 2000-2001 is due to the early withdrawal of monsoon in Rajasthan.
4. Uneven distribution of rains:- Due to comprehensive analysis relating to the fluctuation of rainfall and soil-water regimes of the different regions of the country. It was found that sub humid climatic regions are prone both floods and drought. The annual rainfall of India is about 1200 mm. which is slightly more than the global mean of 900 mm but its distribution is uneven. The regional variation is so large that the khari hills of north east get an annual rainfall greater than 6000 mm where as annual rainfall in Rajasthan desert is even less than 150 mm significant features of the distribution of rainfall is that a major part of rainfall received from S.W. monsoon from June to September. The continuation of this rainfall is quite high or whole of Gujarat and central parts of Rajasthan north of this region the per cent distribution by S.W. Monsoon reduced to 78% in Punjab about 80 – 85% in Haryana and North West Rajasthan there is small contribution (6 – 12%) by winter rainfall.

2. Temperature

Extreme variation in the temperature is the prime characteristic of dryland areas. During the winter season the north-east India experience very cold temperature which range from -9 to 14°C in the Ladakh region, in cold arid zone to 3 – 10°C in arid regions of Punjab, Haryana and Rajasthan frost often occur in north-west India Gujarat and Maharashtra experience minimum temperature of the range from 11 – 15°C while Karnataka and Andhra Pradesh record 16 to -18°C. The lowest temperature is recorded in January in North India and in December in South India, the temperature beings to rise from March onwards. Desert regions of North-west India record the highest day temperature. May and June are the hottest months and the mean annual temperature is about 48°C. On an individual days temperature of the order of 47 to 50°C are also recorded heat waves occurs the maximum temperature in southern part range around 39 – 40°C and may is the hottest month with the onset of monsoon the mean daily temperature fall sharply from 42°C in June to 33°C in July and 36°C in August to September.

3. Sunshine

Duration of hour of bright sunshine is an importance factor in dry areas sky remains generally clear favoring in high in salivation by day and rapid back radiation by night. The daily average sunshine hour are 7 to 9 in winter over hour of India arid zone accept cold arid parts where the duration is less than 2.5 hours per day. During the summer season the arid zones record 9 to 10 hours per day and highest over Rajasthan, Gujarat, Maharashtra regions why in cold arid zone.

It is less than 8 hours with onset monsoon the hours of bright sunshine decreases sharply 4 to 5 per day in southern arid zone 4 to 7 hours in northern arid zone in the month of October. When south west monsoon is withdrawal over major part of the country the highest sunshine bright have 10 per day are recorded over western Rajasthan with describe gradient south wards and are around 5 to 8 hours in southern arid zone on an annual basis the northern add zones recorded slightly higher number sunshine hour per daymore than (>8 hours) as compare to southern arid zone (7 to 8 per day hours).

4. Wind regime

In these areas wind regime as strong as on a Gujarat followed by Western Rajasthan. Where mean wind speed of >20 km/hour during April to July. Mean wind regime is weak in Punjab and Haryana ranging from 2-10 km/hour.

In southern arid zone stronger wind regime (>18 km/hour) prevail in Kurnul region of Andhra Pradesh and over Raichur region Karnataka during June to August the wind regime in general remain low (4 to 10 km/hour) during October to April over major part of southern arid zone.

5. Potential evapotranspiration (PET)

Potential evapotranspiration varies from 1 to 2 cm during the January to 10 to 14 cm during May to August in cold arid zone. The annual PET in this region varies from 75 to 80 cm (6 cm per year) in hot arid zone in north western India the annual PET increase from 130 in Punjab to 160 Haryana to over 200 cm in Rajasthan. During May-June the PET values at α maximum varying from 18-20 cm in Punjab to 22-23 Haryana and 21-31 cm. in Rajashtan during winters. The PET values range between 3 to 8 cm the annual PET values over southern arid zone 180-185 cm because of the proximate to the equitor and the maximum PE values vary farming 22 cm curing April-May while lowest during winter season between 10 to 13 cm.

Other features prevant in rainfed areas are:

1. Prolonged dry spells during the crop period
2. Low moisture retention capacity
3. Poor soil fertility condition
4. Socio-economic constrains particularly because of the pre-dominance of small and marginal farmers 54% of the holdings is less than one hectare.
5. Technological and developmental constraints
6. Limited infrastructure component and improper an untimely availability of credits and agricultural inputs.

Important rainfall features in rainfed areas

The annual average rainfall varies from 400 mm to 700 mm. Year to year fluctuations are so much that there is no guaranteed of fixed quantity of rainfall. Uncertain and ill distributions of rainfall are two qualities which makes the Rainfed farming difficult. Rainfall starts in late June to Early July.

There is depression during late July and early August. Again there is good rainfall in late August and September. The rainfall totally recedes but mid October. The probability of rainfall is more than half of the normal fairly good (P = 0.58) during September.

i) Dry spells

It is another rainfall feature. A break in monsoon is normally experienced (observed) during late July and August. They may extend by 2 week to 13 weeks at a stretch. A break is defined as period receiving less than 15 mm rainfall in consecutive weeks. The normal rainfall during the week being more than 50 mm. Duration of break more than 4 week and frequently more than 3 times usually results in failures crops.

ii) Water availability period

Water availability depends on rainfall and PE. Humid (when rainfall exceeds PE) and moist (when rainfall is less than PE but exceeds PET) period together provides congenial weather for active crop growth.

iii) Wind velocity

Wind velocity is generally high during July and August. If wind velocity exceeds 18-20 km /hr such period coincides with dry spell. Hence evapotranspiration is at high degree. If velocity is low the lowest evaporation rates are observed during November and December.

iv) Bright sunshine hours

Bright sunshine is usually experienced during months of Jan. and Feb. At Solapur it is about 8 to 9 hours. During April and May the sky is usually have with more dust particles, lowest bright sunshine is noticed during Aug. (4 to 5 hours). This indicates the cloudy weather but no rainfall.

v) Humidity

Humidity is high during July and Sept. During Feb. to May it is low. During dry spell, less relative humidity is noticed. Evaporation demands are also accelerated with high temperature and low humidity.

vi) Temperature

Maximum temperature exceeds 41°C during late April and early May. Minimum temperature is noticed during December. Lowest weekly minimum temperature is about 14 to 15 °C. Generally climate is semi and with mild winter and hot summer. Crop like wheat and gram requiring longer cool period hence do poor while prolonged cold weather however, Jowar suffers considerably.

Soil conditions prevalent in rainfed areas

Soils are diverse in their origin, structure, and physical and chemical properties. Important aspects of the formation of soils in dryland environments are the frequently encountered large diurnal changes in temperature that cause mechanical and physical disintegration of rocks and wind-blown sands that score and abrade exposed rock surfaces. The physical disintegration of rocks leaves relatively large fragments which are then slowly broken up by chemical weathering. Vegetation also plays a fundamental role in the process of soil formation by breaking up rock particles and enriching the soil with organic matter from above- and below-ground plant parts. This role can be lessened in dryland environments because of the sparse vegetation and limited development of above-ground plant parts. However, the root systems of these plants are often extensive laterally and vertically extensive and it is this characteristic which mainly affects the soil. Of primary importance to planners and managers of natural and agricultural resources are the water-holding capacity of these soils and their ability to supply nutrients. Soil depth and texture largely governs the amount of water that can be held in a soil body. However, the depth of soils in dryland regions is often limited by a *hardpan layer*, restricting water-holding capacities and rooting depth. These hardpans, often consisting $CaCO_3$ or silicaceous materials, can be more-or-less continuous and occur between 5 and 60 cm below the surface. As there is little deposition, accumulation, or decomposition of organic material in dryland environments, the organic content of the soils is low and, therefore, natural soil fertility is also low. The limited organic matter that is present can be quickly lost when soils are cultivated for agricultural crop production. Soils are often characterized by the extensive leaching of nutrients and intensive weathering of minerals on older exposed surfaces.

The resources at the farm level in rainfed agriculture are land and rainfall. The rainfed areas have a cafeteria of soil situations. They show great diversity in texture, structure, type of clay, organic matter content and depth. These variations induce significant differences in infiltration rate, erodibility, moisture holding capacity, drainage characteristics, aeration, susceptibility to and recovery from compaction and general response to soil management and manipulation. However, only certain soil types occur widely in the seasonally dry tropics.

1. Vertisols

Vertisols (Black Soils) are important both agriculturally and ecologically. According to estimates of the national Bureau of soil survey and land use planning of India, true vertisols and related black soils occupy 72 million ha in India and constitute 22% of the country's geographical area. Vertisols are predominant in the states of Maharashtra, Madhya Pradesh, Gujarat, Andhra Pradesh and Karnataka. Nearly 84% of Maharashtra, 44% of Gujarat, and 38% of Madhya Pradesh are occupied by these soils.

The annual rainfall ranges from 500 mm. These soils have high clay content (30 to 70%) and high water holding capacity, the water intake capacity early in the monsoon season is high dull to deep cracks and larger water retention capacity. The high infiltration rate is further enhanced if the soil management is such that the surface is rough and cloudy and is prepared by the bed and furrow method on a graded contour.

2. Alfisols

Alfisols (Red Soils) are agriculturally important soils found in drylands. The clay is predominantly kaolinitic in nature. This clay is red or reddish brown to yellowish-brown is color. The plant available moisture storage in the root zone of these soils is usually less than 10 cm. These soils occupy about 20% of the geographical area.

These soils are moderately weathered with a bore saturation of about 80% and dominated by calcium. pH of these soil posses an argillic horizon within the profile. The clay content increases with depth although shallow and gravelly soils also occur due to erosion. They also contain distinct layers of gravel and weathered rock fragments at lower depths, often called 'murram'. The rooting depth of crops is limited by presence of such layers or by compact argillic horizon.

Alfisols in these areas are mostly structure less or massive and hence have low hydraulic conductivity. The structures less nature of these soils in due to the low content of fine particles of clay in the surface horizon, predominance of 1:1 type minerals and cropping systems which add only small amounts of decomposed organic matter.

An important effect of lack of aggregation or presence of unstable aggregation is the tendency of the soils to reduce surface roughness, rapidly seal the surface after rainfall and produce crusting with subsequent drying cycles. The water supply of soils is reduced by impaired infiltration due to lower conductivity of the pores wherever vegetative corer is not available to dissipate the energy of falling raindrops. Although there soils are easy to till when wet, they become hard when dry and are difficult to plough.

Cultivation increases infiltration initially but in the long run porosity and infiltration rates is usually lower than in untilled soil under & mulch clover. Under natural vegetation, erosion is very low. But when these soils are denuded and/or cultivated, high rainfall intensities cause particle detachment and degeneration of infiltration capacity. This results in high runoff and soil loss. Furthermore, alfisols generally posses inherently low water-retention characteristic. Crusted Soil surfaces increase the water loss through evaporation.

3. Laterite soils

These soils are well drained with a satisfactory hydraulic conductivity. The clay is mostly either Keolinitic or illitic type. Laterite soils are generally associated with undulating topography in region with a relatively high annual rainfall. These soils cover 13 M ha in India.

4. Alluvial soils

Alluvial soils are generally loamy sands or sandy loams, very deep with low water holding capacity. Water intake is high and the soils tend to crust on the surface with a beating rain. This leads to runoff during subsequent rains. The moisture and drainage characteristic of the soils are highly variable. These soils occupy nearly 21% of the geographical area.

5. Sierozemic soils

Sierozemic soils are extremely high in sands, loamy sands and sandy loams. Calcium carbonate concretions occur at various depths, influencing the effective soil depth. Surface crusting is a serious problem. Soil erosion through wind is common. These soils are light textured and hence hold less water and nutrients. Subsoil salinity is common due to extreme aridity.

6. Submontane soils

The land is generally undulating and the soils are silty loam to loam. They are medium to deep. Land slide and soil erosion are common. The stored water in these soils may vary from 20 to 30 cm per meter profile. These soils are spread in the hill and foothill regions of India. Adoption of appropriate soil and water conservation measure is a must.

Table: Soil, Its Characteristics in tabular form

Soil type	Characteristics
Aridisols	Soil of arid, semiarid and high ground water table area. Clay content-3.5-10%. Poor in nutrient and water holding capacity. pH-7.8-9.2 poor in soil fertility status especially N,P,K,S and Zn Dry for most of the year Salt accumulation at surface resulting in development of salic, gypsic and calcic horizon.
Entisols	Bulk density-1.5-1.8mg/m3. Fe and Al very high upto 40%. CEC–25-40 Cmol/Kg Presence of 1:1 (kaolinite) type clay mineral
Alfisols	Acidis soils (Ph 4.5-6.5) Organic matter (3.0-3.5%) Deficient in p.CEC-10-15 cmol/kg
Mollisols	pH 6.0-7.0 organic matter-2-3% CEC-15-20 Cmol/kg soil

Characteristics of rainfed soils that affect carbon sequestration

Rainfed environments are characterized by a set of features that affect their capacity to sequester C. The main characteristic of rainfed areas is lack of water. This constrains plant productivity severely and therefore affects the accumulation of C in soils. The problem is aggravated because rainfall is not only low but also generally erratic. Therefore, good management of the little available water is essential. In addition, the SOC pool tends to decrease exponentially with temperature. Consequently, soils of drylands contain small amounts of C (between 1 percent and less than 0.5 percent). The SOC pool of soils generally increases with the addition of biomass to soils when the pool has been depleted as a consequence of land uses. Soils in drylands are prone to degradation and desertification, which lead to dramatic reductions in the SOC pool. However, there are also some aspects of dryland soils that work in favour of CS in arid regions. Dry soils are less likely to lose C than wet soils as lack of water limits soil mineralization and therefore the flux of C to the atmosphere. Consequently, the residence time of C in drylands soils is long, sometimes even longer than in forest soils. The issue of permanence of C sequestered is an important one in the formulation of CS projects. Although the rate at which C can be sequestered in these regions is low, it may be cost-effective, particularly taking into account all the side-benefits resulting for soil improvement and restoration. Soil-quality improvement as a consequence of increased soil C will have an important social and economic impact on the livelihood of people living in these areas. Moreover, given the large extent of drylands, there is a great potential for CS. The potential offered by drylands to sequester C is large, not only because of the large extent, but because historically, soils in drylands have lost significant amounts of C and are far from saturation.

Chapter 5

Drought : Types, Effect on Plants, Mechanism of Crop Adaptation under Drought Conditions

Drought

There is no universally accepted definition of drought. Early workers defined drought as prolonged period without rainfall. Drought is a situation when the actual seasonal rainfall is deficient by more than twice the mean deviation. American Meteorological Society defined drought as a period of abnormally dry weather, sufficiently prolonged for lack of water to cause a severe hydrological imbalance in the area affected.

Prolonged deficiency of soil moisture adversely affected crop growth indicating incidence of agriculture drought. It is the result of imbalance between soil moisture and evapotranspiration needs of an area over a fairly long period as to cause damage to standing crops and to reduce the yields.

Important causes for agricultural drought are:

- Inadequate precipitation
- Erratic distribution
- Long dry spells in the monsoon
- Late onset of monsoon and
- Early withdrawal of monsoon

Major droughts in India during last five decades

Year	% area affected in the country	Category
1951	33.2	Moderate
1952	25.8	Slight
1965	42.9	Moderate
1966	32.3	Moderate

1968	20.6	Slight
1969	19.9	Slight
1971	13.3	Slight
1972	44.4	Severe
1974	29.3	Moderate
1979	39.4	Moderate
1982	33.1	Moderate
1985	30.1	Moderate
1986	19.0	Slight
1987	49.2	Severe

Source: Yojna (16-30 June, 1989)

Types of drought

Drought can be classified based on duration, nature of users, time of occurrence and using some specific terms.

Based on duration

a) Permanent drought: This is characteristic of the desert climate where sparse vegetation growing is adapted to drought and agriculture is possible only by irrigation during entire crop season.

b) Seasonal drought: This is found in climates with well defined rainy and dry seasons. Most of the arid and semiarid zones fall in this category. Duration of the crop varieties and planting dates should be such that the growing season should fall within rainy season.

c) Contingent drought: This involves an abnormal failure of rainfall. It may occur almost anywhere especially in most parts of humid or sub humid climates. It is usually brief, irregular and generally affects only a small area.

d) Invisible drought: This can occur even when there is frequent rain in an area. When rainfall is inadequate to meet the evapo-transpiration losses, the result is border line water deficiency in soil resulting in less than optimum yield. This occurs usually in humid regions.

Based on relevance to the users (National Commission on Agriculture, 1976)

a) Meteorological drought: It is defined as a condition, where the annual precipitation is less than the normal over an area for prolonged period (month, season or year).

b) Atmospheric drought: It is due to low air humidity, frequently accompanied by hot dry winds. It may occur even under conditions of adequate available soil moisture. It refers to a condition when plants show wilting symptoms during the hot part of the day when transpiration exceeds absorption temporarily for a short period. When absorption keeps pace with transpiration the plants revive. (Mid day wilt).

c) Hydrological drought: Meteorological drought, when prolonged results in hydrological drought with depletion of surface water and consequent drying of reservoirs, tanks etc. It results in deficiency of water for all sectors using water. This is based on water balance and how it affects irrigation as a whole for bringing crops to maturity.

d) Agricultural drought (soil drought): It is the result of soil moisture stress due to imbalance between available soil moisture and evapotranspiration of a crop. It is usually gradual and progressive. Plants can therefore, adjust at least partly, to the increased soil moisture stress. This situation arises as a consequence of scanty precipitation or its uneven distribution both in space and time. Relevant definition of agricultural drought appears to be a period of dryness during the crop season, sufficiently prolonged to adversely affect the yield. The extent of yield loss depends on the crop growth stage and the degree of stress. It does not begin when the rain ceases, but actually commences only when the plant roots are not able to obtain the soil moisture rapidly enough to replace evapotranspiration losses.

Based on time of occurrence

a) Early season drought**:** It occurs due to delay in onset of monsoon or due to long dry spells after early sowing

b) Mid season drought: Occurs due to long gaps between two successive rains and stored moisture becoming insufficient during the long dry spell.

c) Late season drought**:** Occurs due to early cessation of rainfall and crop water stress at maturity stage.

Other terms to describe drought

a) Relative drought: The drought for one crop may not be a drought situation for another crop. This is due to mismatch between soil moisture condition and crop selection. For eg. A condition may be a drought situation for growing rice, but the same situation may not be a drought for growing groundnut.

b) Physiological drought**:** Refers to a condition where crops are unable to absorb water from soil even when water is available, due to the high osmotic pressure of soil solution due to increased soil concentration, as in saline and alkaline soils. It is not due to deficit of water supply.

General effects of drought

- Delay or prevention of crop establishment.
- Weakening or destruction of established crop.
- Predisposition of crops to insects and diseases.
- Alteration of physiological and biochemical metabolism in plants.
- Alteration of quality of grain, forage, fibre, oil etc.
- Drought is often a factor in yield reduction.

Effect of water deficit on physio-morphological characteristics of plants

Both dryland crops and irrigated crops experience an assortment of ecological or environmental stresses which include abiotic *viz.*, drought, water logging, salinity, extremes of temperature, changes in atmospheric gases and biotic viz., insects, birds, other pests, weeds, pathogens. Dryland crops are more vulnerable to stress in the present context of changing climatic scenarios. The ability to tolerate or adapt effectively by challenging these stresses is a very complicated phenomenon would be due to various plant interactions occurred in the specific environment. Both biotic and abiotic stresses occur at various stages of plant growth and development and frequently more than one stress concomitantly affects the crop. It is difficult to distinguish effects of these stress factors on the performance of crop plants with respect to yield and quality of harvested products. This is of special significance of maximizing productivity of dryland crops in changing climate scenarios, with complex consequences for ecologically and environmentally sound Indian agriculture. In order to successfully meet this challenge, one should understand the various aspects of stresses in view of the current development and to promote a competitive dryland production system.

Abiotic stress

Abiotic stress is defined as the adverse effect of naturally occurring or non-living factors on the living organisms in a specific environment. The non-living variable must influence the environment to adversely affect the population performance or individual physiology of the plant in a significant way. This type of stress is primarily unavoidable and is the most harmful factor concerning the growth and productivity of crops in dryland areas.

Ex. Drought stress, cold stress, salinity, water logging, wind, extremes of temperature, high variability in radiation, changes in atmospheric gases.

Biotic stress

Biotic stress is the stress that occurs as a result of damage done to plants by other living organisms, such as bacteria, viruses, fungi, parasites, beneficial and harmful insects, weeds, and cultivated or native plants. The types of biotic stresses forced on a plant for their growth and development depend on its ability to resist particular stresses under particular environment. Among the biotic stress, the majority of plant diseases are caused by fungi and pests.

(A) Effects of abiotic stress on dryland crops

As said above abiotic stress occurs naturally and agronomists can think of mitigation strategies for this stress. While biotic stress are generally caused by pest and diseases, understanding the physiology is of less importance unless the better productivity is obtained in dryland areas. Hence, more discussion is required on abiotic stress which influencing growth and development of dryland crops in this chapter.

Soil moisture is the most limiting factor for getting profitable yields in dry farming and dryland farming situations. Understanding of the physiological processes that happen during moisture stress is needed to ameliorate the stress effects either by management practices agronomically or by altering the cropping pattern of the regions. Stress is the consequence of action of external factors on an organism/plants. Moisture stress specifies the action of shortage of water or excess of water on plants.

Causes

- Whenever transpiration surpasses absorption of water by the plants leads to stress
- After irrigation or rainfall, water deficit develops due to higher transpiration than absorption especially on hot middays. In such situations, stomata are open for a short period in the morning and evening and partially closed during the rest of the day.
- Whenever soil moisture reaches about -15 bars, plants show wilting symptoms (permanent wilting point) and at -60 bars, plants die permanently (ultimate wilting point).

1. Effect of moisture/drought stress on physiology of plants

a) Water relations

When the moisture is limiting in dryland areas, had negative impacts on water relations in plants such as reduction in leaf water potential and relative water content mainly due to transpirational demands. In addition to this, continuous decline in turgor of the plants likely anticipated. In such plants, leaf wilting is a common symptom could be due to turgor loss. Increase in leaf and canopy temperature might have hindered the plant growth. Turgidity of guard cells decrease with increasing moisture stress in plants.

b) Photosynthesis and respiration

Drought stress induces several changes in various physiological, biochemical, and molecular components of photosynthesis. Photosynthesis is reduced due to reduction in photosynthetic rate, chlorophyll content, leaf area and increase in assimilate saturation in leaves. Biochemical reduction of CO_2 by altering several biochemical process will effect under moisture stress. Reduction in leaf expansion, tillering or branching and increase in leaf senescence in moisture stressed plants is a collective phenomenon. Translocation of assimilates is affected. The photosynthesis apparatus, Photosystem II is comparatively more tolerant to drought stress than heat stress. Respiration increases with mild stress and if the stress continues, water content and respiration will be affected considerably. Moisture stress coupled with high temperature can result in decreases in leaf and root respiration in the short term.

Drought can influence photosynthesis either by stomatal closure and decreasing flow of CO_2 into mesophyll tissue or by directly impairing metabolic activities. The main metabolic changes are declines in regeneration of ribulose bisphosphate (RuBP) carboxylase/oxygenase enzyme. In general, during the initial onset of drought stress, decreased conductance through stomata is the primary cause of decline in photosynthesis. At later stages with increasing severity, drought stress causes tissue dehydration, leading to metabolic impairment.

C_3 plants

C_3 plants (dry land pulses, rice, wheat etc) are closely associated with lower productivity at given water-use than C_4 plants mainly because of RuBP carboxylase which is insensitive to atmospheric CO_2 concentrations. Atmospheric CO_2 concentrations are strongly inhibitive to CO_2 uptake in C_3 plants and fixed directly by RuBP carboxylase. Photosynthesis is smaller in C_3 plants leads to lower crop water use efficiency (WUE) for a given rate of transpiration. Photosynthesis decrease in drought conditions because of stomatal limitations. The drought-sensitive metabolism of the C_3 plants have greater recovery of photosynthesis upon irrigation, through increased stomatal conductance.

C_4 plants

The C_4 plants having "Kranz leaf anatomy" are intimately higher productivity at given water-use. The C_4 crops are maize, sorghum, pearl millet, and various forage grasses is essentially a pumping mechanism that moves CO_2 from the mesophyll cells and causes high CO_2 concentrations in the vascular-bundle sheath cells by using CO_2 from decarboxylated C_4 acids in the mesophyll cells. This results in a high utilization efficiency of low intercellular CO_2 concentrations due to the PEP carboxylase enzyme in the C_4 plant. Photosynthesis is greater in C_4

than in C_3 plants has advantage translated into a greater WUE. Hence, C_4 plants has more drought resistance capacity and resulted in relatively better yield under stress. However, under well-watered conditions C_4 plant had greater WUE leads to better economic returns.

c) Anatomical changes

Drought stress from early growth stages leads to decrease in size of the cells, intercellular spaces, thicker cell wall and greater development of mechanical tissues. Hardening of the cell wall and its reduced extensibility will diminish and even it stops cell growth. Size of epidermal cells are reduced without reduction in the number of stomata per unit number of cells. Stomata per unit leaf area tend to increase under drought stress.

d) Metabolic reactions

Severe water deficit in dryland areas causes decrease in enzymatic activity in the plants. Amount of enzymes involved in hydrolysis remain or increase while peroxidase activity decrease. Accumulation of sugars and amino acids takes place under moisture stress conditions. Proline, an amino acid accumulates under moisture stress.

e) Hormonal relationships

Cytokine, gibberlic acid and IAA, growth promoting hormones produced in the plants were tend to decrease under moisture stressed environments. Contrastingly, growth regulating hormones (ABA, ethylene, betain) will produce and tendency to increase by the moisture stressed plants. ABA known as "stress hormone", acts as water deficit sensor to minimize the loss of tissue water potential. ABA is endogenous hormone also caused dormancy and found in wilting leaves and subsequently induced stomatal closure.

ABA effects and consequences in the plant

- General growth decreases
- Cell division and cell expansion decrease
- Germination and tillering decrease
- Root growth and root hydraulic conductance increase
- Flower abscission increase
- Pollen viability decrease
- Grain and fruit growth decrease
- Starch synthesis in cereal grains increase
- Plant reserve mobilization to the grain increases
- Leaf senescence and dormancy increase

f) Protoplasmic dehydration

Dehydration is severe, protoplasm become rigid and brittle. Tissues become desiccated, protoplasm becomes increasingly dense and its viscosity gradually increases.

g) Growth processes and nutrition in plants

The expansion of cells and cell division are reduced under stress resulted in decreased growth of leaves, stem and fruits. Moisture stress during sowing will severely affect germination and hence leaf area, and leaf expansion and root development. The primary processes involved in the plant growth and development are cell division and cell elongation. Generally, cell division is appraised to be less sensitive to moisture stress than cell enlargement. However, comparatively mild water stress influenced considerably on both cell expansion and cell division, even before photosynthesis or respiration is affected. Under moisture stress conditions, maintenance of cell turgor plays an important role in cell growth. Comparatively, water stress during early stage of the cell division and cell expansion were able to recover fully than at the final phase of cell growth and division. Decreased leaf senescence and loss of leaf area are often termed as tolerance and avoidance mechanism during stress.

- Leaf area expansion is most sensitive growth process and decreased drastically under limited drought stress, mainly because of smaller cells and reduction in the number of cells produced by leaf meristems.
- Leaf area is reduced under stress to moisture. Leaf area is reduced in old leaves by reducing mature cell size, whereas, in younger leaves, inhibition of cell division resulted in fewer cells per leaf.

The general effects

1. Reduction in leaf numbers, rate of expansion, and final leaf size during mild stress.
2. Under severe stress, the rate of leaf elongation decreases and leaf growth can cease.
3. Total leaf area can be decreased under drought stress.
4. Continued drought stress can accelerate leaf senescence and lead to death of leaf tissue, resulting in leaf drop, particularly old and mature leaves.
5. Rewatering plants after a relatively short period of stress (3–5 days) does not completely eliminate the effects of drought on the senescence process. In contrast to drought, heat stress can stimulate cell division and cell elongation rates.

6. Drought stress often decreases stem growth in response to changes in internal water status.

Moisture stress affects fixation, uptake and assimilation of nitrogen in dryland pulses. Nitrogen fixation by leguminous plants reduced due to reduction in leghaemoglobin in nodules and number of nodules. NPK uptake is reduced by moisture stress. Nitrogen assimilation is affected due to reduction in nitrate reductase activity. Nitrogen deficiency under moisture stressed conditions will increase stomatal resistance. Phosphorus deficiency also increases in stressed plants.

h) Developmental and reproductive processes

In general, if stress occurs before flowering, the duration of the crop increases. If the stress occur after flowering, the duration of the crop decreases. Panicle initiation is most sensitive stage to moisture stress due to reduction in cell expansion. Anthesis is another important moisture sensitive stage as moisture stress causes drying of pollen and loss of viability of pollen. Grain development decrease due to reduction in leaf area and photosynthesis.

By the intensity and duration of stress, duration of the developmental stage as well as transition of one developmental stage to other is hampered. Development is mostly influenced by photoperiod (day length) or vernalization is mainly a function of temperature depenedent. Both developmental rate of leaves (individual organs) and the advancement of the entire plant are quantitatively reliant on temperature. For example, high temperature encourages more rapid development of leaf canopy and also causes the overall crop development rate to increase so that the crop growing season is shortened.

Some of the developmental process as influenced by high temperature and subsequently moisture stress are as follows:

1. The length of time from floral initiation (panicle initiation) to anthesis (panicle exsertion) is decreased by moderate drought and/or temperature stress but is increased by severe stress.
2. Drought stress during panicle development inhibits the conversion of vegetative to reproductive phase and plants remain vegetative until the stress is relieved.
3. Panicle initiation in sorghum was delayed by as many as 2-25 days and flowering by 1-59 days under drought stress, with more severe effects when drought was imposed both at early and late stage of panicle development.
4. Drought and heat stress can delay the panicle initiation but also can cause the cessation of panicle development at any stages between panicle initiation and flowering.

5. Severe drought or heat stress inhibits panicle exsertion and also delays flowering.
6. Drought stress or heat stress during flowering and anthesis inhibits pollen development and causes sterility because of decreasing pollen function.
7. Drought and/or heat stress also shortens the spike development duration (period during which potential kernel or seed numbers are determined) and the grain-filling duration (during which the grain or seed weight are determined).
8. Drought stress during later stages of panicle or flower development decreases seed numbers and can also increase the duration from seed-set to full seed growth.
9. For cereal crops, longer periods of vegetative and reproductive development are often necessary to improve reproductive potential (number of productive tillers and kernels) and also leaves and tillers to provide assimilate supply during the grain filling.
10. Both drought and heat stress decreases the seed-filling duration, leading to smaller seed size in rice and legumes such as groundnut, soybean.

2. Effect of heat stress on physiology of plants

At insignificant persistent temperatures, have direct positive influence on rates of development (e.g., seed germination rate and flowering) and occurs most rapidly. At constant soil moisture conditions, percentage seed germination increases with increasing temperature above base temperature and later on decreases at higher temperature. Soil moisture stress coupled with high temperature may greatly affect the growth and development of crop plants in dryland areas. High temperatures generally increase leaf appearance rates and leaf-elongation rates, while decreasing leaf-elongation duration. Heat stress resulted in significant increases in leaf numbers, particularly when reproductive development was arrested without any decrease in leaf photosynthetic rates.

a) Growth and reproductive development

High temperatures at early stages resulted in reduced germination percentage, plant emergence, abnormal seedlings, poor seedling vigor, reduced radicle and plumule growth of geminated seedlings. High temperature causes loss of cell water content for which the cell size and ultimately the growth is reduced. Reduction in net assimilation rate (NAR) is also another reason for reduced relative growth rate (RGR) under high temperature in C_4 crops. The reproductive tissues are the most sensitive to high temperature during flowering time leads to loss of entire grain crop cycles.

- High temperature stress also leads to changes in morphological indications like scorching and sunburns of leaves and twigs, branches and stems, leaf senescence and abscission, shoot and root growth inhibition, fruit discoloration and damage.
- Reduced number of tillers with promoted shoot elongation, leaf area and productive tillers/plant were drastically reduced under HT (30/25 °C, day/night). High temperatures may alter the total phenological duration by reducing the life period.
- Severe heat stress decreases stem growth, decreased plant height, decreased root growth by reduced root number as well as root length and root diameter.
- Significant decrease in floral buds and flowers abortion, in some cases may not produces flowers, fruit or seed.
- High night temperatures (32 °C) increase spikelet sterility resulting from excessive ethylene production was also noticed.

b) Photosynthesis and respiration

Photosynthesis is one of the most heat sensitive physiological processes in plants. The processes of photosynthesis are more tolerant to heat stress and are mostly stable in the temperature range of up to 30 to 35 °C, depending on crop species. However, very high temperatures (>40 °C) can negatively affect photosynthesis. The thermal effects of photosynthesis and respiration are related to membrane function and membrane integrity. In general, heat stress influences membrane fluidity, induces membrane leakiness, and influences the integrity of protein and membranes.

At high temperature, photosynthesis decreases due to decreasing the solubility of oxygen resulting in increased photorespiration. The activity of Rubisco are also decreased at high temperatures by inhibiting the enzyme Rubisco activase. Photosystem II is most sensitive to heat stress and greatly reduced. In general, photosynthesis is temporally (only during daytime) and spatially (only in green tissues) restricted, while respiration occurs continuously and in all organs. Respiration exponentially increases with increasing temperatures from 0 to 35 or 40 °C, reaching plateau at 40 to 50 °C. At temperature above 50°C, respiration decreases because of damage to respiratory mechanism.

- Major changes occur in chloroplasts by causing injury of thylakoids, loss of grana stacking and swelling of grana under heat stress.
- Closure of stomata under heat stress is another reason for decreased photosynthesis that affects the intercellular CO_2.
- Soluble proteins and Rubisco binding proteins (RBP) are reduced under heat stress which resulted in hampered photosynthesis.

- Starch and sucrose synthesis is greatly affected by reduced activity of sucrose phosphate synthase, ADP-glucose pyrophosphorylase, and invertase at high temperature.
- Reduced leaf water potential, reduced leaf area and pre-mature leaf senescence which have negative impacts on total photosynthesis performance of plant under heat stress.
- Accumulation of heat shock proteins (HSPs) is a mechanism of stress resistance to heat which protect the plant tissues.

3. Effects of salinity on plants

Soil salinity is a major factor that limits the yield of agricultural crops, threatening the capability of crop production system. At low salt concentrations, yields are mildly affected or not affected in all the crops. While at high concentrations most of the crop plants, will not grow in high concentrations of sodium salts (100-200 mM NaCl). While the plants which tolerates to high salt concentration (halophytes) mostly grown in coastal and arid regions can survive in excess of 300-400 mM due to specific mechanisms of phylogenetic adaptation.

High salinity affects plants in the following ways:

- High concentrations of salts in the soil disturb the capacity of roots to extract water (osmotic effect), and
- High concentrations of salts within the plant itself can be toxic due to inhibition of many physiological and biochemical processes such as nutrient uptake and assimilation and hence growth of plants reduced.
- Old leaves in the sensitive plant die and reduce the photosynthetic capacity of the plant. This exerts an additional effect on growth.
- Water use by plants is decreased due to reduced shoot growth by reduction in the leaf area development relative to root growth.

Prominent changes occur in plants under salinity/sodicity stress are mentioned below:

1. High Na+ transport to shoot
2. Preferential accumulation of Na in older leaves
3. High Cl^- uptake by the plants
4. Lower K+ uptake by the plants
5. Lower fresh and dry weight of shoots and roots
6. Higher partitioning of dry matter to sensing leaves
7. Low P and Zn uptake
8. Poor root growth.

Synthesis of Heat Shock Proteins (HSPs) under high temperature stress

Synthesis of HSPs or stress proteins, is a tolerance mechanism occuring in plants. Synthesis of HSPs is extensive response to diversity of environmental stress such as drought, salinity, heavy metal stress and oxidative stress. These are highly conserved and produced in all plants to protect from the damage caused in plants from high temperature/heat stress as well as other stresses. Heat shock proteins acts as a defence mechanism against high temperature that leads to changes in the cellular membrane structure, protein metabolism, level of enzymes and photosynthesis activity. Hence these are called "Heat-shock proteins or Stress-induced proteins or Stress proteins". Heat stress/high temperature, affects the metabolism and structure of plants, especially cell membranes and many basic physiological processes such as photosynthesis, respiration, and water relations.

Effects of biotic stress on dryland crops

Biotic stress is stress that occurs as a result of damage through to plants by other living organisms viz., bacteria, viruses, fungi, parasites, beneficial and harmful insects and weeds. The difference between biotic and abiotic stresses are discussed earlier. Among the biotic stress, the major stress which occurred economic loss of plants under dryland regions are pest and diseases and the intensity and occurrence depends on climate and topography. Losses caused by biotic stress remains high and farmers may face greater difficulty in controlling biotic stresses compared to abiotic stress. The biotic stress injury mainly influence on crop yields that influences population dynamics, plant development and ecosystem nutrient cycling.

Losses due to pests and disease in crop plants remain to pose a significant threat to crop production and ultimately food security. Now a days, intensive farming systems became increasingly dependent on synthetic chemical pesticides for controlling pests and diseases and became unsustainable. Application of more and more pesticides leads to development of resistance in the target pests, and has negative impacts on biodiversity. More importantly, plants will have increased susceptibility to pest due to the implications of climate change. Under such situations, gaining better knowledge on physiology of plants could leads to sustainable control of biotic stress. For plant biotic stress resistance and process to manipulate, we require a detailed knowledge of these interactions at a wide range of scales.

Effect of biotic stress on plant growth

Both insect and diseases will affect photosynthesis, as chewing insects reduce leaf area and virus infections reduce the rate of photosynthesis per leaf area.

Vascular- wilt fungi greatly affects the water transport and photosynthesis by inducing stomata closure in the plants.

Response of plants to biotic stress

Selection of wide range of plants and suitable crop varieties defences against pathogens and pests which act to minimize frequency and impact of attack. It include both physical and chemical adaptations. Utilization of high metal ion concentrations derived from the soil permit plants to reduce the harmful effects of biotic stressors (pathogens, herbivores *etc.*) called induced resistance. Meanwhile preventing the severe metal toxicity by way of safeguarding metal ion distribution throughout the plant with protective physiological pathways. At the same time, successful pests and pathogens have evolved mechanisms to overcome both constitutive and induced resistance in their particular host species.

Mitigation of abiotic stress by crop management

Apart from adoption and resistance mechanisms, we need to know other mitigation strategies for maximizing productivity and profitability of the dryland farmers sustainably by preserving resource base of the dryland soils.

1. Conservation tillage

The conservation tillage system is designed to conserve soil moisture from one season to another or from one year to the next. Conserving soil moisture is virtuously, depending on climate and crops of the regions. Increasing storage of soil moisture by with or without conservation tillage is typical moisture conservation practice in dryland farming. The benefit of conservation tillage in terms of increasing available soil moisture to the crop depends on soil water-holding capacity, climate, topography and management practices. Percent increase in soil moisture availability to the crop but the amount of soil moisture is not remarkable under probability of crop failure and success. The conservation tillage carries additional benefits such as improved soil nutrients availability and the eradication of certain soil-born pests, such as nematodes.

Conservation tillage is not a novel concept or practice, which has recently gained wider and sometimes enthusiastic acceptance. The principles involve, minimized tillage operations to conserve soil structure and to maintain ground cover by mulch, such as stubbles which reduce water runoff and increase soil infiltration. Rotation of crops under dryland areas is not suitable because of soil moisture constraints. Intercropping with cover crops/ pulses could increase the soil moisture availability by harvesting runoff and utilized for next season crop.

Deep tillage is a system to overcome hardpan, very high bulk density and compacted soils can be performed by deep plowing or deep ripping. Deep plowing involves actual plowing to depth which is an expensive operation and is uncommon in dryland farming. Deep ripping is somewhat less expensive and more often used in crop production. The important consideration in deep ripping is to operate at the correct depth in order to break the hardpan, no less and no more.

2. Furrow ditches/dikes

These techniques constitute a field surface tillage manipulation to minimize runoff away from the field and provides more time to infiltration. Short basins were made as furrows by small dikes and is very amenable to grow row crops such as cotton, corn and sorghum. It is generally considered effective for increasing rainfall capture and raising dryland crop yields where annual rainfall ranges between 500 and 800 mm.

3. Water harvesting and spreading

As discussed earlier water harvesting is broad technique to conserve moisture in dry land areas in minimizing abiotic stress.

4. Diversification of farming

Diversification of farming is an ancient but an effective approach to reduce the risk associated with farming in unpredictable environments. Reduced diversification to the extent of mono-cropping is possible only with a high level of control like irrigation over the crop environmental conditions. Diversification of cropping to reduce risk is especially important under dryland conditions. Crop diversification takes an advantage where crops grown in a single stress environment. Crops differ in their response to a given environment and this difference is used to reduce the risk associated with growing one crop. "Mixed cropping" or "intercropping" is an example of a traditional and a successful approach to crop diversification on a single piece of land, where two or more crops are grown together in various possible patterns. Planting of various crop varieties offer a better chance of reducing loss due to environmental stress, as compared with growing one variety only. Varietal diversification is constructed mostly on differential phenology, primarily flowering dates when two or more environmental stress likely occurs.

Climatological approach for crop planning

Climatological analysis helps to identify cultivars suitable for different regions. Water balance for different agroclimatic regions has been calculated and water availability periods worked out. The detailed information on the above said aspects has been described in the tables given below:

Table: Length of effective cropping season in different areas of India

Category	Effective cropping season (weeks) in different areas			Potential cropping system
< 20 weeks	Bellary (8)	Jodhpur (11)	Anantpur (13)	Sole cropping
	Hisar (17)	Rajkot (17)	Bijapur (17)	
20-30 weeks	Jhansi (21)	Kovillpatti (21)	Hydrabad (22)	Intercropping
	Udaipur (22)	Solapur (23)	Agra (24)	
	Anand (25)	Akola(27)		
>30 weeks	Bhuneshwar (32)	Varansi (32)	Indore (36)	Sequence cropping
	Banglore (36)	Hoshiarpur (35)	Dehradun (44)	
	Rewa (36)	Ranchi (45)		
	Samba (44)			

Table: Suitable cropping systems based on rainfall and water availability period

Rainfall (mm)	Soils	Water availability period (Weeks)	Potential cropping systems
350-600	Alfisols and shallow vertisols	20	Single kharif cropping
350-600	Aridisols and Entisols	20	Single cropping either in kharif or Rabi
350-600	Deep vertisols	20	Single rabi cropping
600-750	Alfisols, vertisols, Entisols	20-30	Intercropping
750-900	Entisols, deep vertisols,		
	Alfisols and Inceptisols	>30	Double cropping with monitoring
>900	Entisols, deep vertisols, Alfisols and Inceptisols	>30	Assured double cropping

Different climatic conditions are observed in dryland areas which are given below:

1. Variation in onset.
2. Dry and wet spells.
3. High temperature.
4. Low humidity.
5. Soil erosion.

1. Variation in onset of monsoon

In dryland areas the onset of monsoon at different time creates the main problem in crop cultivation.

a) Delayed onset of monsoon.

b) Good start of rains followed by dry spells.

c) Early withdrawal of monsoon.

d) Extended monsoon.

a) Delayed onset of monsoon

Delay in onset of monsoon is a common feature of monsoon with delay ranging from 15 days to 50 days. Under delayed monsoon black cotton soils (<45 cm) if diverted for rabi sorghum, production of kharif crops especially pulses and oilseeds is affected as well as the yield of rabi jowar results into poor returns.

b) Good start of rains followed by dry spells

This situation is very common. Usually one or two dry spells are experienced during Kharif season. The dry spells may be short of 5-7 days to long with more than 15 days. Under these conditions following measures may be adopted for Kharif crops.

i) Controlling plant population for conserving available moisture.

ii) Checking weed growth to reduce moisture loss.

iii) Increasing interculture practices to prevent evaporation.

iv) Choice of crops like redgram which can sustain longer breaks.

c) Early withdrawal of monsoon

This situation creates two problems in rabi.

i) Sowing of rabi crops may be suspended.

ii) When crop is sown, it requires moisture conservation practices e.g. reduced plant density, use of surface mulch, protective irrigation, stripping of leaves.

d) Extended monsoon

Such situation is rarely experienced. Double cropping is possible in medium deep soils. Sowing of rabi crops are extended. In certain areas' sorghum may be replaced by gram and wheat as the sorghum may suffer due to cold temperatures in winter.

2. Dry and wet spells

The rainfall is mainly in the form of wet spells with dry spells extending from few days to many weeks.

3. High temperature

Due to low rainfall in dryland areas, deficiency of water in soil occurs. In dryland areas the evaporation rate and transpiration rate become high. Soil become dry due to which the growth of plants is affected. High temperature affect the seed germination, emergence of the seedling, respiration rate, metabolic process, development of grains, pollination fertility and total yield of the crop. Due to high temperature the transpiration rate of plant leaves increases day by day. Leaves colour become yellow.

4. Low humidity

Due to low rainfall intensity, the humidity of the atmosphere decrease which affect overall crop production.

5. Soil erosion

Due to low rainfall the soil erosion increases. Due to soil erosion the fertility of soil decreases thereby many nutrients which are present in soil moves from one place to another place. Due to soil erosion disturbance in soil structure, fertility of soil decrease, water holding capacity becomes poor and poor stand of plants in soil.

Crop planning for erratic and aberrant weather conditions

1. Choice of crops

Crops and varieties earlier grown on dryland under acute drought conditions were not necessarily efficient with regard to moisture use. Screening for efficient crops and crop varieties began in 1965 when hybrids and high yielding, short duration and photo insensitive varieties appeared. Efficient crops and varieties for each of the agro-climatic zones have been identified. Specific crops have been screened according to the severity of drought which are mentioned hereunder

a) For Very Acute Drought **:** Pearlmillet, Groundnut, sesame, cotton, Bengal gram (Chickpea), Barley, mustard, cluster bean, castor, moongbean, urdbean and cowpea.

b) For Acute Drought **:** Sorghum, wheat groundnut, ragi, barley, Bengalgram, mustard, toria, taramira, safflower. Pearlmillet was found to be the best crop for acute drought conditions. (Also includes the crops recommended for very acute drought).

c) **For Moderate Drought :** Wheat, Bengalgram, Barley, Rice, Maize, Mustard and Pigeonpea. (Also includes the crops recommended for very acute and acute drought conditions).

2. Choice of suitable crop varieties

For general recommendation, the drought resistant crops and varieties are –

Pearlmillet	:	MBH-110, BK-560, MBH-118, WCC-75 and ICNS-7703
Sorghum	:	N-35-1, GSH-1 and Swarna
Maize	:	Pusa ageti-76, Diara Composite, Ganga-5 and Vikram
Castor	:	Aruna, Bhagya and Soubhagya
Cotton	:	H_4, PRS-72 and Bikaneri Lerma
Groundnut	:	Kauvery-3, NH-2 and JL-24
Pigeonpea	:	Pusa ageti, AS-5 and AS-8.
Barley	:	Ratna, DL-3, K-226 and R.D. 31
Gram	:	Pusa-209, H-208, K-468, Pent 114 and G-235
Wheat	:	C-306, WL-410, Kundan, K-65, K-68 and Meghdoot.

1. Intercropping

Intercropping is a practice most common is dryland areas. It acts primarily as an insurance against failure of crop in the event of aberrant weather.

Advantages

Intercropping is done mainly for better use of resources, less incidence of pests, disease or weeds and improved nitrogen economy when legume is an intercrop. Intercropping does not cause so much stress on soil moisture as does single cropping. This is because plants of the same species compete for water more vigorously with each other than to the plants of different species. In intercropping the root systems, growth habits and energy requirement period of the component crops differ, so that the stress on moisture is divided in amount and time. Intercropping of pearlmillet with either green gram or cowpea, resulted in a significant increase in the grain yield of pearlmillet over the sole crop of pearlmillet either as uniform row system or as paired row system with some yield of intercrops also. Thus, intercropping is a best proposition under acute drought conditions.

New technological modifications in time, technique and pattern of planting of crops grown in association, have evolved intercropping as economically viable and a feasible practices.

a) Planting pattern

The experience gained at several research centers with various crops has shown that the paired row system of planting enables adoption of more efficient system of intercropping on drylands. Paired row system of planting gave higher yield, more favorable microclimate for plant growth and development, besides, efficient water use and is the principal reason for superior performance of paired row planting over uniform row planting. Paired row system of planting increased the utilization of moisture from deeper layers. Paired row planting has lesser leaf area index (LAI) than regular row planting to reduce evapotranspiration and conserve moisture and nutrients. It maintained a mild crop canopy microclimate which reduced or suppressed weeds by 50-72% in different crops. Yield under paired row system was comparatively more than the uniform row system of planting.

b) Planting time

Under acute drought conditions, sowing of crops largely depends upon the onset of monsoon rains. Under acute drought conditions, it is better to sow the crop at onset of first monsoon rain.

2. Fertilizer management

Fertilizer use mainly depends upon nutrient status and soil-moisture availability. So, under acute drought conditions, fertilizer application should be adjusted according to the moisture regime. Simultaneously, nutrient deficiency reduces water use efficiency (WUE) even under limited moisture. Therefore, a moderate amount may increase WUE. However, on arid lands it is extremely difficult to adjust fertilizer a given crop should be decided on the basis of the water storage capacity of soil, plant density and water habit of the crop. The new genotypes (recently released hybrids) are found to be more efficient in total uptake of nutrients and transfer of nutrients from vegetative parts to grain.

a) Dose of fertilizer

Fertilizer application under acute drought conditions should be adjusted according to the soil moisture regime. A moderate amount of suitable fertilizers in accordance with the soil moisture may increase water use efficiency (WUE). Under dryland conditions found that fertilizer levels and volume of water increased the grain yield and net return significantly with the application of 10,000 litres of water/ha even without addition of fertilizers. The yield of grain increased significantly when the crop was fertilized with 20 kg P_2O_5/ha and a very high magnitude was obtained when this dose of fertilizer was supplemented with 20,000 litres of water/ha at sowing time.

b) Method of fertilizer application

The method of fertilizer application has an important bearing on fertilizer use efficiency (FUE) under acute drought conditions. FUE can be increased by placement of fertilizers.

c) Time of fertilizer application

The time of N application depends on the uptake requirement of the plant and moisture availability. Traditionally, pearlmillet is grown in light sandy loam type of soils. Therefore, high intensity of rainfall at early stage of crop growth caused N loss due to leaching of NO_3-N and thus the low availability of N appreciably reduced the later growth and development. Therefore, it was recommended to apply N fertilizer in splits. The general recommendations are ½ of total N to be applied at planting and the other half after 3 weeks of planting or half N at planting, one fourth after 2 weeks and rest ¼ N after 6 weeks of planting.

Application of farm yard manure (FYM) has also its significance under acute drought conditions. FYM acts as much besides providing the major nutrients to the crop plant FYM acts as surface barrier to the evaporation losses from soil which consequently increases the water use efficiency (WUE).

3. Use of biofertilizers

Nitrogen and Phosphorus are the key element for crop production and productivity. But in the present era of energy crisis, chemical fertilizers, being non-renewable energy sources, are not only costly but also inaccessible to the poor farmers. Therefore, it is now imperative to explore the renewable sources of fertilizers such as bio-fertilizers as substitutes for chemical fertilizers, which not only increase the productivity but also improve the soil fertility. The seed treatment with bio-fertilizers before sowing when grown under dryland conditions has been a common practice and proved to be extremely beneficial. *Azospirillum* and V.A.M. (*Vesicular Arbuscular Mvcorrhiza*) are the two important micro-organisms being evaluated intensively at present as bio-fertilizers for supplying N and P, respectively throughout the world.

4. Contingent crop planning for different aberrant weather conditions

In the post following four types of weather aberrant conditions were experienced, which result in occurrence of acute drought conditions, resulting in low productivity of different dryland crops. The contingent crop planning for each and every type of aberrant weathers have been evolved which are described as :

a) Normal onset of monsoon followed by long gap in rainfall

i) **Selection of Crops and Varieties:** In case of very early break in monsoon i.e. if drought is experienced 10-15 days after seeding and seedlings are killed, the crops like bajra (MBH-110), Jowar (CSH-6) and Castor (GAUCH-1) are resown, which serve as mid season corrections.

ii) If the drought occurs about 30-45 days after seeding, gap filling/ transplanting in case of pearlmillet, followed by light top dressing of 10-15 kg N/ha, serves as a good measure.

iii) Inspite of sufficient initial rainfall, the crop suffers when there are long gaps in subsequent rains. To tide over such adverse situations, the crops should have a good, deep root system. When the crop is given adequate fertilizers, it develops a deeper and denser root system and extracts moisture from deeper soil layers.

b) Delayed onset of monsoon

i) Choice of proper crops and varieties

a) When there is failure of rains upto late July the Crops Varieties suggested are bajra (MBH-110), Jowar (CSH-6), Castor (GAUCH-1) and Sesamum (Purva 1).

b) If the rains are delayed beyond end of July i.e. upto mid of August, then it will be advisable to sow bajra (MBH-110), Castor (GAUCH-1) and jowar for fodder.

ii) Transplanting of pearlmillet

If the rains are delayed beyond the above periods and start somewhere during end of August and growing season is reduced to 60-70 days, transplanting of pearlmillet seedlings followed practice to overcome this problem. Three weeks old pearlmillet seedling can be transplanted successfully from end of July to end of August.

c) Early withdrawal of rains towards the end of season

i) Harvesting of *Kharif* crops for fodder only if the rains stop very early i.e. by the end of August or Ist week of September.

ii) Harvesting *Kharif* crops at physiological maturity and preparing field for *rabi* crops.

iii) Application of surface mulches for conservation of residual soil moisture.

iv) Adjustment of plant population: Plant population may be adjusted accordingly soil moisture availability by thinning the plants within the row and in extreme cases by removing the plants within the row and in extreme cases by removing third row or alternate rows. In the intercropping system, the fodder component may be harvested if the drought prolongs.

v) **Protective irrigation:** Recycling of harvested water at the critical growth stages during drought periods could save the crop from the complete failure of crops.

d) Total failure of monsoon rain with scattered rainfall

Such condition of acute drought was experienced during 1987. The contingent crop planning for such conditions are :

i) Selection of suitable crops and varieties**:** Pearlmillet with its variety MBH-110 proved to be the best measure to tackle such conditions and only short duration crops like grain legumes (Black gram and green gram) should be grown.

ii) Spray of Kaolin (6% suspension) on crop plants as light reflectant to reduce transpiration losses from the plants.

iii) Hoeing or interculture operations between the rows of crops to break the crust to conserve soil-moisture by breaking the capillaries by creating a fine "dust mulch".

5. Alternate crop strategies

Sowing of crops under drylands depends upon the rains. Often the rains may commence quite early or may be delayed considerably and the some crop may not be sown over this entire period. Yields may be reduced considerably because of the incidence of pests and disease or due to a shorter growing season. By an intense comparison of crops sown on different dates, all the research centers could identify the most suitable crop at given point in time.

6. Mid-season corrections

If the acute drought occurs after the crop is growing in the field, there are certain options that can be employed, such as thinning, ratooning resowing/ replanting, when moisture conditions are again favorable. If the drought occurs during very early stage of crop i.e. within 10-15 days after sowing, it is better to resow the suitable crop and variety with subsequent rains. On the other hand, if the drought is experienced when the crop is of 40-50 days, other means like thinning or ratooning should be considered.

Plant population may be adjusted according to soil moisture availability by thinning the plants within the row or in extreme cases by removing third row or alternate row if the crops experience the acute drought when it is about 40-50 days old.

Under acute drought conditions, thinning of plants by removing every third row was found to be advantageous to mitigate moisture stress.

7. Crop life-saving measures

The source of soil moisture is ultimately the rainfall and in the tropics, the rainfall is limited to a few months with unpredictable distribution. Any attempts to store the moisture either for extended use or use at later dates when the crop experiences the acute drought conditions, should be considered as a crop life-saving measure which includes :

A. Run-off harvesting (Water Harvesting)

B. Life-Saving Irrigation or Run-off recycling

C. Use of mulches

D. Weed Control

E. Off-season tillage

(A) Run-off harvesting (water harvesting)

Moisture stress is the most important constraint to crop production in drylands of India. In many dry parts of Western Rajasthan (Jaisalmer, Bikaner, Ganganagar and parts of Jodhpur), where local precipitation is the only source of soil moisture for crop production but alone is not sufficient to ensure a crop, harvested water plus rain water may suffice the crop needs. Thus rain water conservation under dryland agriculture can be made possible by adopting water harvesting techniques

For water harvesting efforts should be made to increase the runoff and reduce infiltration. This harvested water may be used as "life-saving" irrigation when the crops experience an acute drought. There are two main types of water-harvesting :

1. *In Situ* water harvesting or field water harvesting or Micro water harvesting.
2. Catchment water harvesting or collection of run-off in Drought ponds for recycling.

1. *In situ* water harvesting

In situ water harvesting is the best way of storing rainwater for crops grown under acute drought conditions. It creates micro-watersheds within a more or less leveled field. It can be achieved by land configuration into ridges and furrows on grades formed prior to sowing or with eventual cultural operations. The crop is sown in narrow strips between vide intervals that are ridged as artificial miniature watersheds. The run-off slopes are compacted and hardened to reduce infiltration and induce run-off to crop rows. The usual ratios between the relative width of run-off strips and cropping belts are from 2:1 to 4:1 depending upon the amount of rainfall than can be expected. Generally, two main approaches are involved in preparing the runoff strips :

i) Slopes covered with plastic films, rubber or metal sheets. (But the main disadvantages of plastic coverings are their high costs and its susceptibility to damage by high winds).

ii) Water proofing and surface stabilizing by spraying adhesives.

There are two methods of field water harvesting :

a) Inter-plot water harvesting **:** Different rations of catchment to cropped area, namely, 1:1 with slope on one side 1:1 with slopes on both sides and 1:2 with slope only on one side, were studied but catchment to cropped area ration 1:1 with slopes on both sides led to move uniform water harvesting, efficient use of moisture and ultimately higher yields

b) Inter-row water harvesting**:** Among the systems of water harvesting, inter-row water harvesting appears to be more practical, feasible and acceptable system to the farmers, for little land is sacrificed for the purpose.

Other examples of in situ water harvesting are – graded border strips-dead furrows etc. The most suitable land configuration systems found suitable for different zones suggested by Venkateswarlu (1983) are :

2. Catchment water harvesting

Construction of a farm pond is another way to collect excess run-off water from a given catchment area, which can be used as "life-saving" irrigation. The pond water is subjected to less through deep percolation in the soil profile and evaporation to the atmosphere. The percolation/seepage losses are high in sandy and sandy loam soil (20-40%), therefore, without the use of an effective sealant, the water will not be available for more than 3-4 weeks.

(B) "Life-Saving" irrigation or runoff recycling

The next crucial question is how to use the harvested water effectively, efficiently and profitably. This harvested water can be used to irrigate the crop as "life-saving" irrigation when the crop experience acute drought. Studies conducted at the dry land Projects show that about one ton of a additional cereal grains can be harvested by adding 5 ha cm of water. This benefits in terms of increased of harvested water to dryland crops.

(C) Tillage, weed control and intercultural operations

Attempts should be made to increase the infiltration and to reduce run-off under dryland conditions. Tillage operations like ploughing and chiseling open up the hard layers of soil and increase infiltration. Sub soiling gave the maximum water use efficiency (WUE) followed by chisel ploughing, mould board ploughing, local ploughing and surface cultivation.

Under acute moisture stress it is desirable to control the weeds because, the weeds are highly wasteful of water. Weeds compete the crop plants not only for limited moisture supply but also for nutrients, light, CO_2 and space. Weed damage is more at early stage of crop grow.

Now the pre-emergence application of Atrazine @ 0.5 kg/ha plus one hoeing at 6 week stage of crop controlled the weeds most effectively and resulted in a significant increase in grain yield as well as not return over unweeded control. This treatment is also effective for conserving moisture because atrazine is also an anti-transpirant and hand hoeing creases a fine "dust mulch" on soil surface by breaking the crust and capillaries –

(D) Use of mulches, antitranspirants and other measures

It is well known that soil moisture from sub-soil layers rises to the surface and is lost as evaporation. Simultaneously, the loss as transpiration from plant surface is also of significant importance. To control evaporation losses, use of mulch is good. For transpiration control, spraying of antitranspirants and light reflectant is done. Under acute drought conditions, topping and seed hardening is also done.

Different dry farming treatments on pearlmillet are topping upper one third plants within 30 days of sowing, spraying of borax (0.2%), kaolin (6%) and atrazine (100 ppm) on plants, pre-sowing seed hardening with 0.8% KNO_3 and 0.2 NaCl and applying of dry surface mulch, resulted in a significant increase in grain yield over untreated control pre-sowing seed hardening with 0.2% KNO_3 resulted in maximum grain yield of 23.7 q/ha of pearlmilet. But, seed hardening is only practices when the drought is during early season. If the acute drought is experienced during the crop growth period, the best preposition is to apply the dry surface mulch (straw mulch) @ 5 t/ha in between the rows of standing crop.

Mulching to conserve soil moisture by breaking the capillaries by creating a fine "dust mulch".

8. Alternate crop strategies

Sowing of crops under dryland depends upon the rains. Often the rains may commence quite early or may be delayed considerably and some crops may not be sown over this entire period. Yields may be reduced considerably because of the incidence of pests and disease or due to a shorter growing season. By an intense comparison of crops sown on different dates, all the research centres could identify the most suitable crop at given point in time. Such information is useful for farmers whose resources are meagre and could not sow the crops in time.

Mechanism of crop adaptations under moisture deficit conditions

Levitt (1972) distinguishes between two basic ways in which plants can grow and survive in dry habitat: (a) escaping drought, and (b) actual drought resistance.

Escaping drought

Evading the period of drought is the simplest means of adaptation of plants to dry conditions. Many desert plants, the so-called ephemerals germinate at the beginning of the rainy season and have an extremely short growing period, which is confined to the rainy period. Between germination and seed maturity, as few as five (Polunin, 1960) to six (Kassas, 1966) weeks may suffice. These plants have no mechanism for overcoming moisture stress are therefore not drought resistant

The presence of germination inhibitors in such seeds serves as an internal raingauge, which permits germination only after an amount of precipitation has fallen that will be sufficient to remove the inhibitor (Went, 1965).

Early maturity

In cultivated crops, the ability of a cultivar to achieve maturity before the soil dries out is the main adaptation to growth in a dry region. However, only very few crops have such a short growing season, that they can be considered similar to the 'ephemerals' in escaping drought. One example is of certain varieties of millets that can produce mature seeds within 60 days from germination.

Drought resistance

Plants that cannot escape periods of drought can adapt to these conditions essentially in two ways:

Avoiding stress

Stress avoidance is the ability to maintain a favourable water balance and turgidity even when exposed to drought conditions, thereby avoiding stress and its consequences. Stress avoidance is mainly due to morphological-anatomical characteristics, which themselves are the consequence of the physiological process induced by drought (Levitt, 1972). These xeromorphic characteristics are quantitative and may vary according to the environment conditions

A favorable water balance under drought conditions can be achieved either by: (i) conserving water, by restricting transpiration before or as soon as stress is experienced (the so-called water savers); or, (ii) accelerating water uptake sufficiently, so as to replenish the lost water (water spenders).

The mechanisms for conserving water :

1. Stomatol mechanisms: Stomata of different species vary widely in their normal behaviour and range from staying open continuously to remaining closed continuously. Many cereals open their stomata only during a short time in the early morning and remain closed during the rest of the day. There are however differences in this respect between varieties of the same crop, as shown by the following examples : in two varieties of oats, the one more resistant to drought opened its stomata more rapidly in the early morning when moisture stress is at its minimum and photosynthesis can proceed with the least loss of water (Stocker, 1960). An interesting observation made in the course of trials on the effects of soil moisture regime on different wheat varieties was that the stomata of a semidwarf variety remained open throughout the day, while those of the tall variety were open for only a few morning hours remaining closed thereafter even under favourable conditions of soil moisture and transpiration (Shimshi & Ephrat, 1970).

   However, mechanisms of conserving water based on the closure of stomata will inevitably lead to reduced photosynthesis and may lead to "drought induced starvation injury" (Levitt, 1972).

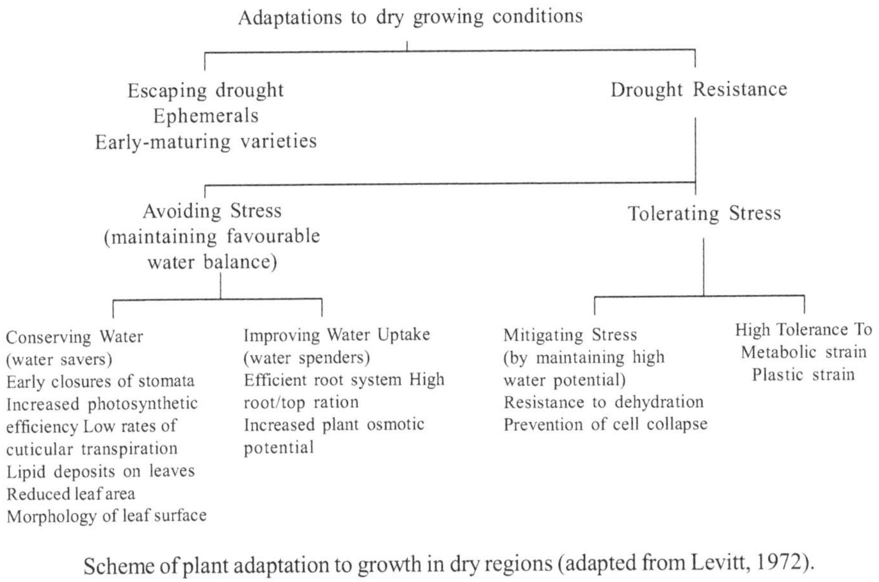

Scheme of plant adaptation to growth in dry regions (adapted from Levitt, 1972).

2. Increased photosynthetic efficiency: One possibility for overcoming the limitations on photosynthesis imposed by stomatal closure as a means for increasing resistance to loss of water by transpiration would be a higher rate of CO_2 assimilation for a given stomatal opening (Hatch & Slack, 1970). A number of important crop plants (maize, sugarcane, sorghum, proso, foxtail and finger millets) (Hatch *et al.*, 1967) as well as certain forage species, Bermuda grass (*Gynodon dactylon*), Sudan grass, Bahia grass (*Paspalum notatum*), Rhodes grass, (*Chloris gayana*) (Murata & Iyama, 1963) and certain *Atriplex* spp., fix most of the CO_2 into the C_4 of malic and aspartic acids; the so-called C_4-dicarboxylic acid (C_4) pathway. They have a primary carboxylating enzyme which has both a high potential activity and a high affective for CO_2. with such an enzyme, high rates of fixation of CO_2 and with relatively restricted stomatal opening (Loomis *et al.*, 1971). The species using the C_4 path say have a high rate of CO_2 assimilation for a given stomatal opening, higher temperature and light optima for photosynthesis (El-Sharkawy *et al.*, 1968), and higher light saturated rates of apparent photosynthesis (El-Sharkawy & Hesketh, 1965).

   Another photosynthetic process which reduces water loss without a concomitant reduction in photosynthesis is found in a number of succulent plants, which keep their stomata closed during most of the day. These plants have the so-called Grassulacean acid metabolism (GAM) which enables them to fix large amounts of CO_2 as organic acids at night and convert it into carbohydrate during the day. As the primary assimilation of CO_2 occurs at night, the lower temperatures prevailing limit the amount of water transpired for a given stomatal opening. The pineapple, for example, which has the GAM type of photosynthesis, produces about the same amount of dry matter per year as sugarcane, but often uses only 10 to 20% as much water. With the exception of the pineapple and certain *Agave* spp. grown for fibre and paper pulp, GAM plants have not been exploited agriculturally (Loomis *et al.*, 1971).

3. Low rates of cuticular transpiration: The typical example is the cactus.

4. Decreasing transpiration by a deposit of lipids on the surface of the leaves on exposure to moderate drought. This has been shown to occur is soybeans (Levitt, 1972).

5. Reduced Leaf area: The principal means of reducing water loss of xeromorphic plants is their ability to reduce their transpiring surface. Apart from the common means to keeping the aerial parts small, perhaps the simplest form of this reduction of the transpiring surface is the rolling or curling of leaves at times of water stress, a characteristic phenomenon exhibited by many grasses.

The rolling of leaves has been shown to reduce transpiration by almost 55 per cent in semiarid conditions, any by 75 per cent in desert xerophytes (Stalfelt, 1956).

6. Leaf surface: Various morphological characteristics of leaves help to reduce the transpiration rate and may affect survival of plants under drought conditions leaves with thick cuticle, a waxy surface and with the presence of spines, etc., are common and effective. In certain species, drought stimulates the production of epidermal hairs (Shields, 1958). The mere presence of pubescence on the leaves is no longer considered as an advantage in reducing transpiration (Kramer, 1959), unless it be thought increasing their albedo.

7. Stomatal frequency and location: A smaller number of stomata can retard the development of water deficits. In certain species, the stomata are located in depression or cavities in the leave, which is a feature that can further reduce transpiration by limiting the impingement of air currents.

8. Effect of awns : Awned varieties of wheat predominate in the drier and warmer regions, and have been found to yield better than awnless ones, especially under drought conditions, though there are exceptions (Grundbacher, 1963). Awns have chloroplasts and stomata and so can photosynthesize; it has been found that the contribution of the awns to the total dry weight of the kernels was 12% of that by the entire plant. (Me-Donough & Gauch, 1959).

Accelerating water uptake :

1. Efficient root systems: The root systems of drought resistant plants are characterized by a wide variety of apparent adaptations. These are responses to such predominant soil conditions as the duration of soil dryness and the depth that is normally wetter, the presence and nature of soil constituents and the degree of soil salinity. Plants become adapted to dry condition; mainly by developing an extensive root system, rather than by structural modifications of the roots (Shields, 1958). The concept "extensive root system" also includes, in addition to growth in depth and lateral growth, the density of roots per unit volume of soil and the number of secondary hair roots (Oppenheimer, 1960).

   There are considerable differences between cultivated plants in the extent, depth, and efficient, of their root system. Sorghum has nearly twice as many small branch roots per unit length of main roots as has maize and it is assumed that this is one of the main reasons for its greater drought resistance. Another adaptation of some xerophytes is their speedy response to a rewetting of the soil by the rapid formation of secondary rootlets ('rain roots') which may appear only two to three hours after a fall of rain (Evenari, 1962). An important characteristic of certain cultivated plants is their ability to respond quickly to fluctuations in moisture content by having roots that grow rapidly towards sources of available moisture. It has been shown experimentally that drought resistance of varieties of spring wheat is related to the rapidity of growth of the primary roots (Aamodt & Johnston, 1936).

Plants are also known to differ with respect to the ability of their root systems to penetrate relatively dry soils. Oats, wheat and barley show little penetration of the soil, at or below wilting point, while two range grasses – Side-oats grama (*Bauteloua curtipendula*) and Love-grass (*Eragrostis eragrostis*) penetrated the soil quite extensively (Salim *et al.*, 1965).

2. High root-to-top ration (RIT*)*: A high root-to-top ration is a very effective means of adaptation of plants to dry conditions (Killian & Lemee, 1956), as under such conditions the growth rate of the roots considerably exceeds that of the shoots. The transpiring surface is thereby reduced, while the root system of the individual plant obtains its water from a large volume of soil. An increased root : top ration may actually result in a greater amount of total dry weight of plants grown under dry conditions as compared to similar ones grown with full moisture.
3. Differences in osmotic potential of the plants: Levitt (1958) has calculated that a difference of 0.5% in the soil moisture content that induces permanent wilting, could supply a plant with enough water to keep it alive for 6 days, this could mean, in certain cases the difference between survival and death.
4. Conversion of water spenders to water savers: Because of their increased water absorption, water spenders are characterized by very high rates of transpiration. However, as soon as the absorption rate becomes insufficient to keep us with water loss, the water spenders generally develop some of the characteristics of the ware savers (Levitt, 1972).

Drought tolerance

When the plant is actually submitted to low water potentials it can show drought tolerance by either mitigating the actual stress induced by the moisture deficiencies, or by showing a high degree of tolerance to these stresses.

Mitigating stress: Adaptations to drought based on mitigating the effects of stress permit the plants to maintain a high internal water potential in spite of drought conditions. They therefore are able to maintain cell turgor and growth, avoid secondary drought induced stress, as well as direct or indirect metabolic injury due to dehydration (Levitt, 1972).

1. Resistance to dehydration: The simplest method of avoiding drought induced damage is by resisting dehydration – preferably to the extent of maintaining

turgor, and at least by avoiding cell collapse after loss of turgor (Levitt, 1972).

2. Prevention of leaf collapse: In many instances the principal effect of leaf adaptation to dry conditions, such as thick cell walls, is to prevent wilting or collapse of the leaf, rather than to ensure reduced water loss. This explains why xeromorphic types of leaves frequently transpire more, when amply supplied with water, than mesomorphic types (Shields, 1958; Kramer, 1959), and thus make possible an efficient use of water during the short periods of wet conditions which are typical for dry environments. As soon as the soil moisture-level again becomes critical, the stomata close and the morphological adaptations to reduced water supply mentioned above become operative.

Tolerating stress :

1. Resistance to metabolic strain: The greater the elastic dehydration strain [Cell dehydration is elastic and completely reversible up to a point, beyond which it is plastic, irreversible and therefore injurious (Levitt, 1972)], the greater the danger of a resulting plastic metabolic strain. When dehydration occurs, there are mechanisms for tolerating an accompanying plastic metabolic strain.

   Starvation : The greater the dehydration a plant can tolerate and still keep its stomata open the lower will be its dehydration compensation point [This is the water potential at which rates of photosynthetic CO_2 absorption and respiratory CO_2 evolution are equal, so that net photosynthesis equals zero]. This kind of drought tolerance will be due to the specific ability of the fuard cells to remain turgid even when the leaf cells are wilted, and is possible due to an increase in the solutes of the guard cells (Levitt, 1972).

   Protein loss: The avoidance of protein loss under drought conditions is the net result of two processes : a decreased rate of protein breakdown or and increased rate of synthesis. Young leaves are more resistant to drought than older leaves and this is ascribed to their greater protein (Chen *et al.*, 1968). In tests with varieties of winter wheat, for example, water stress caused a greater decreased in RNA content in a nonresistant variety than in the resistant variety (Stutte & Todd, 1967).

2. Avoidance or tolerance of direct drought-induced plastic strain: Some higher plants may show little restriction of transpiration during drought and can become air dry without suffering permanent injury. Grama grass, for example, may lose 98.3% of its free water without injury (Oppenheimer, 1960). Exposure to a sublethal stress may cause an increase in resistance so that the plant is able to survive an otherwise lethal stress. This has led to the developing of methods for increasing hardiness of crop plants.

Drought management strategies

In our country, irrigated area constitutes 33% and 67% is dryland and rainfed, out of 143.8mha of cultivated land. Total cropped area is 180.36 m ha. Such dryland agriculture contributes about 45% of national food grain production. After full exploitation of dryland, it may contribute up to 75% of total food grain production. Pulses and oilseeds are mainly grown in such areas. Important commercial crops viz. cotton, castor groundnut and all coarse grains viz. jowar, bajra, maize crops are rainfed. Problems of dryland agriculture are inadequate, uncertain, uneven rainfall. Drought is major problem, so we go for contingency planning.

Nature of drought

The different strategies for drought management are discussed under the following heads.

- Adjusting the plant population: The plant population should be lesser in dryland conditions than under irrigated conditions. The rectangular type of planting pattern should always be followed under dryland conditions. Under dryland conditions whenever moisture stress occurs due to prolonged dry spells, under limited moisture supply the adjustment of plant population can be done by
- Increasing the inter row distance: By adjusting more number of plants within the row and increasing the distance between the rows reduces competition during any part of the growing period of the crop. Hence it is more suitable for limited moisture supply conditions.
- Increasing the intra row distance: Here the distance between plants is increased by which plants grow luxuriantly from the beginning. There will be competition for moisture during the reproductive period of the crop. Hence it is less advantageous as compared to above under limited moisture supply.

Mid contingent plan for aberrant weather conditions

The contingent management practices done in the standing crop to overcome the unfavorable soil moisture conditions due to prolonged dry spells are known as mid season conditions.

- Thinning: This can be done by removing every alternate row or every third row which will save the crop from failure by reducing the competition
- Spraying: In crops like groundnut, castor, red gram, etc., during prolonged dry spells the crop can saved by spraying water at weekly intervals or 2 per cent urea at week to 10 days interval.

- Ratooning: In crops like sorghum and bajra, ratooning can practiced as mid season correction measure after break of dry spell.

Mulching: It is a practice of spreading any covering material on soil surface to reduce evaporation losses. The mulches will prolong the moisture availability in the soil and save the crop during drought conditions.

Weed control: Weeds compete with crop for different growth resources ore seriously under dryland conditions. The water requirement of most of the weeds is more than the crop plants. Hence they compete more for soil moisture. Therefore the weed control especially during early stages of crop growth reduces the impact of dry spell by soil moisture conservation.

Water harvesting and life saving irrigation: The collection of runoff water during peak periods of rainfall and storing in different structures is known as water harvesting. The stored water can be used for giving the life saving irrigation during prolonged dry spells.

Short term strategy of contingency planning

Immediate planning

- Judicious use of surface and groundwater for drinking and irrigation and districts.
- Ensuring availability of quality fodder to animals.
- Livestock management including establishment of fodder/feed depots and cattle camps.
- Selection of crops, cropping sequences and agronomic practices for drought affected areas.
- Promotion of subsidiary income and employment generating activities.
- Gainful implementation of NREGA, RKVY, NFSM, NHM, RGGVY, BRGF and other schemes.
- Deployment of Information Technologies.

Contingent cropping

- Date bound contingency crop plan for different meteorological sub-divisions and agro-ecological regions.
- Early maturing varieties of different Kharif crops.
- Crops and cropping system-wise agronomic practices.
- Technology for Rabi/summer Rice in Boro cultivation area.
- Potential districts for Rabi Maize.
- Soil & Water Conservation Measures for different rainfall regions.

Promotion of subsidiary income and employment generating activities

- Extraction of gum from arid land trees and bushes such as *Acacia senegal.*
- Collection of *Prosopis juliflora pods* and its post harvest processing as animal feed and human food.
- Mushroom cultivation, bee keeping, sericulture, tasar cultivation etc.
- Salt making from saline ground water.
- Commercial rising of the nursery for trees, vegetables and annual flowers.
- Multiplication of root stocks as well as nursery of fruits and flowers.

Compensatory production for kharif deficit

- Boro Rice
- Winter Maize
- Wheat and other Rabi Crops
- Intensification of Rabi pulses and oil seeds

Medium and long term strategy

- Securing good quality water in drought prone areas.
- Perennial and non-conventional fodder.
- Improved live-stocking, breeding and management.
- Up gradation and fine tuning of crops, cropping and farming systems.
- Exploiting under-exploited and under-utilized plant resources.
- Creation of alternate income and employment generating opportunities.

Major policy issues

- Land related policies.
- Water related policies.
- Other policies:
- Feed, fodder and seed banks
- State Level Policy for Livestock.
- Inclusion of coarse cereals in PDS and procurement programme.
- A unique 4-5 years rolling system of credit and repayment.
- Value addition and marketing.
- Strengthening of power distribution grids for hiked demand.
- Augmented supplies of diesels.

Contingent plan for rabi

- Time bound Rabi Plan based on receipt of rainfall in August and September.
- Resource conservation technologies including zero tilled specially in Indo-gangetic plains.
- Need based location specific agronomic manipulations.
- Resorting to cultivation of less water demanding crops and cropping systems in limited irrigation water supply areas.

Preparation of appropriate crop plans for dryland areas

Although, the past meteorological records help sufficiently in planning for future, but because, the drought is a phenomenon which may occur at any time during the crop period, so, we have to make our planning accordingly. All India Coordinated Research Project for Dryland Agriculture has evolved several contingent crop production strategy to meet the challenges of acute drought conditions, which are discussed here under :

Modern practices

- Choice of suitable crops
- Choice of suitable crop varieties
- Cropping system (Intercropping) with
- Planting pattern
- Planting time
- Fertilizer management with dose, time and method of application.
- Seed hardening

a) Mid season corrections

- Alternate crop strategy
- Transplanting of pearl millet
- "Life-saving" irrigation
- Adjustment of plant population
- Leaf stripping

b) Moisture conservation practices

- Use of mulches
- Use of antitranspirants and light reflectants

- Interculture operation
- Weeding
- Water harvesting techniques
- Traditional Practices :
- Contour building
- Ridge and furrow system
- Deep ploughing

At the household level, the strategies adopted before drought include:

- Crop diversification/ crop variety diversification
- Livelihood diversification
- Staggering of seedling periods
- Early migration
- Keeping land fallow
- Limited re-sowing.

Strategies adopted after drought sets in include:

- Increased re-sowing if the rains come
- Shift to pulses on upland and medium-level land
- Early sowing of rabi crop
- Providing irrigation to paddy seedlings
- Transplantation of over-mature paddy seedlings to get straw for feeding animals
- Migration
- Asset depletion
- Borrowings
- Reduction in consumption

In fragile zones and among the poorest households, reduction in consumption is ultimate coping mechanism.

Agronomic practices for acute drought conditions

Although, the past meteorological records help sufficiently in planning for future, but because, the drought is a phenomenon which may occur at any time during the crop period, so, we have to make our planning accordingly. All India

Coordinated Research Project for Dryland Agriculture has evolved several contingent crop production strategy to meet the challenges of acute drought conditions, which are discussed here under –

Modern practices

- Choice of suitable crops
- Choice of suitable crop varieties
- Cropping system (Intercropping) with
- Planting pattern
- Planting time
- Fertilizer management with dose, time and method of application.
- Seed hardening

Mid season corrections

- Alternate crop strategy
- Transplanting of pearlmillet
- "Life-saving" irrigation
- Adjustment of plant population
- Leaf stripping

Moisture conservation practices

- Use of mulches
- Use of antitranspirants and light reflectants
- Interculture operation
- Weeding
- Water harvesting techniques
- Traditional practices :
- Contour building
- Ridge and furrow system
- Deep ploughing

Chapter 6

Water Harvesting: Importance, its Techniques, Efficient Utilization of Water through Soil and Crop Management Practices, Study of Mulches and Antitranspirants

Natural resources conservation and their management hold key to sustainable agriculture and livestock production. It is all the more crucial for countries with predominant agrarian economies where development of sustainable agriculture is essential for overall growth, redressal of poverty and security. Conservation of both soil and rain water as very crucial and basic resources have been practiced since ancient times in India. However, there has been renewed emphasis on conservation and efficient utilization of these resources in the recent past. The fact that 70% of the arable area of India is rainfed with precarious supply of water and that rainfed agriculture supports 44% of India' s human population and contributes 90 per cent of coarse cereals and pulses, 80 per cent of oilseeds and 65 per cent of cotton and growing realization that further gains i n productivity of crops and livestock will emanate from rainfed regions, leave no room for complacency in this regard.

Year after year, the fate of a vast 'majority of Indian farmers hangs in balance, as success with rainfed agriculture continues to be a gamble. It is evident that crop yields in semi-arid areas depended more on rainfall distribution than on total rainfall and lack of serious efforts to create d water supply for crops through scientific management of rainwater is a factor favouring this avoidable uncertainty. In rainfed agriculture no other input can perhaps enhance the yield without effectively tackling of the rainfall aberration related sub-optimal moisture availability. Therefore a prerequisite for substantial improvement in the agriculture production in the semi-arid region is to manage runoff water and to use it either at the time of moisture stress even during the monsoon or in next season. It has been reported that only one supplementary irrigation at proper stage can double of Rabi crops. But supplemental water is a developed resource and is more expensive than the natural resources. Hence, it is all the more necessary to use this water judiciously and efficiently. Standardization of techniques by which as much of precipitation as possible can be conserved for crop use, either directly in the soil profile through infiltration or through runoff collection and recycling is an area of priority in research.

Rainwater availability in India

On an average, India receives an annual rainfall of 1,100 mm, which ranges between 100 mm in Western India (Thar desert) to over 10,000 mm in the northeastern parts. India can be divided into four major zones based on the average rainfall.

Measured in terms of irrigation water equivalence, India annually receives 400 million hectare meters (M ha m) of rainwater. India's share thus works out to about 4% of the global precipitation (11 billion ha m). In fact, she receives the largest amount of rainwater among the countries comparable to her size. The southwest monsoon (June-September rains) contributes 74% of the total precipitation. In contrast, the northeast monsoon provides 30%. The remaining 23% is supplied during pre- (~I 0%) or post-monsoon (-13%) seasons.

Out of 400 M ha m rain water being received in the country, 178 M ha is available as surface flows. Big reservoirs can at best store about 70 M ha in. Nearly 24 M ha, in surface flow is possible to stash at the primary source or the donor areas. Approximately 6.32 M ha in is harvestable through field level structures in low to medium rainfall regions (~1000 mm and less), where certainty in agricultural productivity is more often uncertain.

Water harvesting and recycling

It involves collecting excess runoff, natural or induced, in surface reservoir for agricultural use. Water thus harvested can be stored in surface reservoirs or ponds or in soil. Appropriate water harvesting strategies depend on soil type, terrain, and rainfall characteristic.

Water harvesting

'Water harvesting' is generally used as an umbrella term covering a range of methods of collecting and conserving various forms of runoff from different sources, Water harvesting can be done both *'in situ'* and dugouts /ponds depending upon the situation as well as socioeconomic status/capabilities of the fanner. In dryland agriculture, water harvesting usually denotes the collection of excess run off in a storage tank and using it for the batterment of crop production in the collected or other areas.

The water collection in the farm pond is directly used for protective irrigation. The water stored in other structures will recharge the ground water and used for protective or supplementary irrigation. Collecting and storing water on the surface of the soil for subsequent use is known as water harvesting. It is a method to induce, collect, store and, conserve local surface runoff for-agriculture in arid and semiarid regions. Water harvesting is done both in arid and semiarid regions with certain differences. In and regions, the collecting area or catchments area is substantially in higher proportion compared to command area. Actually, the runoff is induced in catchment area in and lands whereas in semiarid regions, runoff is not induced in catchment area, only the excess rainfall is collected and stored. However, several methods of water harvesting are used both in arid and semiarid regions.

Runoff from cultivated fields should be led through grassed waterways to suitably located excavated or embankment farm ponds. The runoff harvesting, storage and utilization involves -

a) Selection of catchment;

b) Selection of site and estimation of size of farm pond;

c) Construction of farm pond including inlet;

d) Laying of silt trap and spillways;

e) Minimization of seepage, percolation and evaporation losses;

f) Utilization of stored water for life saving irrigation of kharif crops or for establishment of rabi crops in the catchment.

A) Water harvesting *In situ*

Water harvesting techniques adopted *'in situ'* are inexpensive to augment rain water use efficiency in drylands, any process of reducing the runoff volume will increase moisture infiltration in the soil leading to increased moisture volume.

'In situ' water harvesting and run off recycling are the potent measures of crop life saving during periods of moisture stress. The method of *'in situ'* water harvesting helps in improving the crop stand, especially on small seeded crops like pearl millet by mitigating the adverse crust effects on germination and seedling burying problems due to rain during germination. This practice also helps in safe disposal of excess rainwater in high rainfall areas. Appropriate land configuration for *'in situ'* moisture conservation are broad-beds and furrows, graded border, strips, inter-row and inter-plot water harvesting systems. For shallow alfisols and related soils even dead furrows serve this purpose very well. Strategies for 'in situ' moisture conservation involve contour farming and installing barriers on the contour.

a) Contour Farming **:** Contouring or contour farming implies performing all farm operations on the contour rather than up and down the slope. Contour farming is one of the easiest and most effective and low cost method of controlling erosion and conserving moisture. With contour farming, tillage operations are carried out along contours. It creates numerous ridges and furrows, which increase surface detention capacity, increase time for water to infiltrate, and decrease runoff amount, harvesting sizable amount of run off inside the field itself. Contour farming is however, effective on gentle slopes of up to 5%, steeper slopes require slope management techniques. Contour farming can be sometimes inconvenient because it possibly involves frequent turning of farm vehicles and loss of area, which has to be put into buffer strips.

b) Strip Cropping : Another aspect of contour farming is strip cropping or growing crops into long narrow strips established on the contour. In this system, open row crops (e.g., corn) are grown in alternate strips with close canopy crops (e.g., soybeans, alfalfa, etc.). The close canopy crop is often grown in a contour strip down slope from the open row crop. It is also important to establish buffer strips on the contour. There are various types of strip cropping. On the basis of objectives, vegetative materials used, and field design adopted, buffer strips are called contour strip cropping, buffer strip cropping, barrier strips, border strips, or field strips. In addition to reducing runoff amount, establishing strips against the prevailing wind direction also decreases wind erosion.

c) Live beds : One or two live beds (2-3 m wide) can be created either on contour or on grade in the inter bund space. The vegetation on the beds may be according to the liking of cultivator; it can be annual or perennial or a combination of both.

d) Vegetative hedges : Vegetative hedges are established on the contour to create a barrier and increase time for water to soak into the soil. Vegetative hedges are mostly established from bench-type grasses. A widely adapted grass for tropical eco-regions for vegetative hedges is Vetiver *(Vetiveria zizanoides)* or khus grass. Vetiver is densely tufted bunch grass, which can be easily established. In addition to grasses, vegetative hedges can also be established from woody perennials, e.g., *Leucaena leucocephala, Gliricidia sepium,* etc. Properly established and adequately maintained. Vegetative hedges decrease runoff velocity, encourage sedimentation, and reduce runoff and soil erosion. However, closely spaced and narrow strips of grass or woody perennials are likely to be more effective in reducing runoff and soil erosion than narrow or single-row hedges.

e) Tied ridging : This is a modified version of ridge-furrow system where ridges are "tied" at 15 to 20 m interval to allow rainwater collection in the furrows and its infiltration into the root zone. The system not only conserves rainwater but also substantially reduces sediment and nutrient losses induced by runoff.

f) Broad bed and furrows : Broad bed and furrow system implies shaping alternate bed and furrows. This technique is especially suited to black soils, where crops are sown on preformed beds. This system is made before the season and is maintained year after year. The planting is done on the bed. Generally, the depth of each furrow is kept 0.15 m and the inter furrow spacing is maintained at 1.5 m.

g) Dead furrows : Dead furrows are laid across the land slope in rolling lands, to intercept the run-off. The spacing between dead furrows varies between 2 to 5 m or 4 to 7 crop rows. This system works well in alfisols.

h) Bunds : These are low height earthen embankments constructed across the slope in cultivated lands after deciding location of waterways. The bunds function to intercepting runoff, increase infiltration opportunity time and dispose excess rainfall safely. Such bunds can be either contour bunds or graded bunds; many-a- time, these bunds are adjusted with field boundaries, if deviation from grade or contour is not too much; no doubt, spacing depends on many engineering considerations.

i) Contour bunds: are recommended in dry farming areas with light textured soils of slopes up to 6 per cent and where annual rainfall does not exceed 600 mm. They are designed for an expected run-off of 24 hours duration and 10 years frequency. Surplus arrangements (waste weirs) are provided to dispose of the excess run-off beyond the design storage. The cross section area of contour bunds follows the depth and type of soil; however, 0.5 m 2 is minimum.

A small section bund may be created across the slope at half of the vertical bund spacing. Such bunds be nearly 0.1 m2 in section; may be renovated at an interval of 2 to 3 years.

j) Graded bunds : These are constructed in medium to high rainfall areas in permeable soils (vertisols), having 2 to 6 per cent slope. They are also quite suitable for the soils having crust formation tendency like red 'chalka' soils of Telengana region of Andhra Pradesh. By and large; graded bunds with 0.3 to 0.5 m^2 section are constructed with longitudinal gradient of 0.2 to 0.4 per cent depending on the site condition. Graded bunds particularly are made with the following objectives:

i) To reduce run-off

ii) To reduce soil loss; and

iii) To divert run-off

Thus, graded bunds along with waterways and water harvesting structures not only check soil erosion, but they also provide an ideal rainwater management system for many watershed situations.

k) Compartmental bunds : Compartmental bunds, converting the area into square/rectangular pareels - are useful for temporary impounding of water for improving moisture status of the soil. These are made using bund formers. In medium deep black soil, they are found advantageous in storing the rainfall received during the rainy season in the soil profile - there- by augmenting the soil moisture for use by rabi crops. The size of the compartments may be fixed.

   Compartmental bunds are usually practiced in deep black soils with relatively flat terrain and under low to moderate rainfall. The field is laid out into compartments of 6m x 6m to 10 m x 10 m using **a bund former.** The harvested water in these compartments conveniently infiltrates into the root zone and conserved in situ.

l) Terracing : Cultivated lands having land slopes above 10 per cent particularly in hilly areas - should be put under bench terracing by converting the lands into series of platforms. The width of bench terrace depends on the land slope and the permissible depth of cut. Bench terracing is very much effective in reducing soil erosion in hilly areas. At Ootacamund, it has helped in bringing down the annual erosion rate from 39 t/ha to less than 1.0 t/ha on 25 per cent sloping lands. At places where scattered stones are available, loose stone walls can be made to act as risers for bench terrace construction.

B) Run-off harvesting

In rainfed areas, one or two occasional heavy rainfall occurs which creates run off. In vertisols, the runoff is about 60 % and in alfisol it is about 40 %. Hence, efforts should be made for proper utilization of the run off after taking measures for *'in situ'* soil moisture conservation.

Methods of run-off harvesting

The different methods of water harvesting that are followed in and and semiarid regions are discussed separately.

I) Arid regions

The catchment area should provide enough water to stature the crop and the type of farming practiced must make the best use of water. In general, perennial crops are suitable as they have deep root systems that can use runoff water stored deep in the soil, which is not lost through evaporation.

Runoff farming

Ancient runoff farms in the Negev desert in Israel had several cultivated fields fed by water from watersheds of 10 to 50 hectares. The watersheds were divided into small catchment areas of one to three hectares that allowed runoff water to be collected in easily constructed channels on the hillsides' and were small enough to prevent uncontrollable amounts of water. The channels that led the water to cultivated fields were terraced and had stone spillways so that surplus water in one field could be led to lower ones. Farmers constructed small check dams with rocks across the small gullies and guided the water to fields.

Water spreading

In arid areas, the limited rainfall is received as short intense stones. Water swiftly drains into gullies and then flows towards the sea. Water, is lost to the region and floods caused by this sudden runoff can be devastating often to areas otherwise untouched by the storm. Water spreading is a simple irrigation method for use in such a situation. Flow waters are deliberately diverted from their natural courses and spread over adjacent plains. The water is diverted or retarded by ditches, dikes, small dams or brush fences. The wet flood plains or valley floods are thereby used to grow crops.

Micro catchments

A plant can grow in a region with too little rainfall for its survival if a rainwater catchment basin is built around it. Micro catchments used in the Negev desert range from 16 m^2. Each is surrounded by dirt wall of 15.20 cm height. At the lowest point within each microcatchment, a basin is dug about 40 cm deep and a tree is planted in it. The basin stores the runoff from micro catchment.

II) Semi-arid regions

Water harvesting techniques followed in semiarid areas are numerous and also ancient. The objectives of water harvesting structures are:

(i) To store water for supplemental and off-season irrigation,

(ii) To act as silt detention structure;

(iii) To recharge the ground water;

(iv) To raise aquaculture species; and

(v) Recreation and allied agricultural uses.

Thus in most situations, in addition to reviving traditional water harvesting systems, other measures are also needed. Some of such typical water harvesting structures adopted in watershed is as follows:

Dug wells

Hand dug wells have been used to collect and store underground water and this water is lifted for irrigation. The quality of water is generally poor due to dissolved salts.

Tanks

Runoff water from hillsides and forests is collected on the plains in tanks. The traditional tank system has following components viz. catchment area, storage tank, tank bund, sluice, spillway and command area. The runoff water from catchment area is collected and stored in storage tank on the plains with the help of a bund. To avoid the breaching of tank bund, spillways are provided at one or both the ends of the tank bund to dispose of excess water. The sluice is provided in the central area of the tank bund to allow controlled flow of water into the command area. The command area of many tank ranges from 25 to 100 hectares. In areas receiving annual rainfall of about 1000 to 1500 mm with runoff around 50 per cent, where canal and well irrigation is not feasible due to topography and underground water table, tanks are found suitable. An earthen dam of 12 mt in height stores about 55,600 litres of water from a forest watershed of I 0 hectares. T his tank can provide supplemental irrigation for 20 hectares of rainfed farm land. However, there are a few tanks with command area of more than 1,000 hectares. Unlike wells, the quality of water is good in tanks. Water from the tanks is used to irrigate the command area by gravity flow.

Percolation tanks

Flowing rivulets or big gullies are obstructed and water is ponded. Water from the ponds percolates into the soil and raises the water table of the region. The improved water level in the wells surrounding the percolation tanks is used for supplemental Irrigation.

Minor irrigation tanks

Minor irrigation tanks are constructed across the major streams with canal system for irrigation purpose by constructing low earthen dams. A narrow gorge should be preferred for making the dam in order to keep the ratio of earthwork to storage as minimum. The height of the dam may vary from 5 to 15 m. The tanks are provided with well-designed regular and emergency spillways for safety against side cutting. In micro-watersheds, water harvesting *bundhies,* similar to small scale irrigation tanks are also recommended. By and large, the water harvesting *bundhies* are not integrated with extensive canal system.

Nala bunds and percolation tank

Nala bunds and percolation tanks are located in the nalas having permeable formations with the primary objective of recharging the ground water. A strict regulation on the silt load entering the downstream reservoirs is an additional advantage of percolation tanks. The percolation tanks encourage the digging up of wells downstream of recharged area for irrigation purposes. The percolation tanks; are provided with emergency spillways for safe disposal of flow during floods.

Stop dams

Stop dams are permanent engineering structures constructed for raising the water level in the nala for the purpose of providing life saving irrigation during drought periods. These are located over flat nalas at narrow gorges carrying high discharge of long durations. They are created over stable foundation conditions where hard murrum or rocks are encountered. For the stop dam, a site with larger water storage capacity would be desirable.

Farm ponds

Farm ponds are typical water harvesting structures constructed by raising a n embankment across the flow direction or by excavating a pit or a combination of both. Dug out ponds in light soil require lining of the sides as well as the bottom with suitable sealants. This is not the case with heavy black soil. Considering the benefit of the harvested water, different type of linings, viz. brick, concrete, HDPE, etc. may be used.

Types of farm pond

Farm ponds are small storage structures used for collecting and storing runoff water. As per the method of construction and their suitability for different topographic conditions farm ponds could be classified into three categories :

1. Excavated farm ponds suited for flat topography,
2. Embankment ponds for hilly and rugged terrains with frequent wide and deep watercourses, and
3. Excavated-cum-embankment type ponds for soils having mild to moderate sloping topography

Excavated farm ponds are of three types according to their shape. They are:

1. Square
2. Rectangular
3. Circular ponds

Among these, the circular ponds have the geometrical advantage that they have the highest storage capacity and have least circumferential length for a given surface area and side slopes. However, their curved shape is disadvantageous in as much as substantial area is normally lost for agricultural operations. Hence, either square or rectangular ponds are suggested.

Selecting a catchment for a farm ponds

Selection of catchment for locating a farm pond is extremely important for several reasons. It is selected on the basis of its potentiality for yielding sizeable quantity of runoff. Too big a catchment results in a rapid silting, while too small a catchment may not bring in enough water into the pond. The runoff amount from different catchment will depend on several factors tabulated below:

Factors governing amount of runoff

S. No.	Major factor	Associated factors
1.	Rainfall	1. Rainfall intensity
		2. Rainfall duration
		3. Rainfall distribution
		4. Events of rainfall causing runoff.
2.	Land topography	1. Degree of land slope
		2. Length of run
		3. Size and shape of the catchment
		4. Extent of depressions and undulations on the catchment.
3.	Soil type	1. Soil infiltration rate
		2. Antecedent soil moisture
		3. Soil texture
		4. Soil structure
		5. Soil erodibility characteristics
4.	Land use pattern	1. Cultivated or uncultivated or partially cultivated.
		2. Under pasture or forests
		3. Bare fallow or with vegetation
		4. Soil and moisture conservation measures adopted or not.
		5. Crop cultural practices that are adopted.

II. Ridge and furrow system

The system envisages planting of upland rainy season crops on 15-20 cm high ridges with 0.5 per cent grade. The system facilitates root zone aeration besides ensuring disposal of runoff at safe velocities.

III. Broad bed and furrow system

The system involves creation of 90-150 cm wide, 15-20 cm high raised beds with 0.3 to 0.5 per cent grade. The beds are separated by 50 cm wide furrows that drain into grassed waterways. The beds are stable for 2 to 4 years and conveniently adapted to planting of upland crops in rows spaced at 30, 45, 75 or 150 cm. The system tends to conserve soil, rainwater in situ and improves crop yields and sustainability.

IV. Graded bunds

Graded bunds are constructed to control soil erosion on lands having slope in the range of 1 to 6 per cent, These are constructed across the slope on a grade of 0.1 to 0.2 per cent spaced at suitable vertical intervals, so as to divide total catchment area into small parts and runoff water from each part is carried away along these bunds without causing any erosion. Diversion of runoff water coming off a hillock into cropped fields saves the land from degradation. This can be achieved by constructing a bund with a shaped drainage channel along it on a grade of 0.2 to 0.3 per cent. The cross-section or the bund will depend upon the rate of runoff and catchment area contributing runoff to be diverted.

V. Raised and sunken bed system

The system consists of an array of raised and sunken beds of 6-8 m and 3-4 m width, respectively with elevation difference of 15 to 30 cm. The system is created by mechanically shifting soil from demarcated 3-4 m wide strips, designated as sunken beds to adjoining 6-8 m wide strips called raised beds. Sunken beds are tied with small cross section earthen bunds of about 10 cm height at 20 m interval to ensure uniform spread of runoff in sunken beds, Raised beds are planted to an upland crop whereas rice is planted in sunken beds. Raised and sunken bed system facilitates drainage of upland crops, encourages *'in situ'* rainwater conservation and minimizes soil erosion and nutrient losses. The runoff from raised beds is arrested in the adjacent sunken beds. The specification of land configuration depends on amount and intensity of rainfall, runoff potential and water intake rate of soil.

Recycling of water, harvested and stored in farm ponds

Runoff collected into farm ponds can be used in the following situations.

a) Saving a standing kharif crop from drought effect.

b) Providing pre-sowing irrigation for rabi crop.

c) Extending the growing season for the benefit of long duration crops (castor) and

d) Providing a minimal irrigation for growing vegetable/fruit or fodder trees in small areas.

The harvested water needs, to be judiciously used for crop life saving/supplemental irrigation to increase and stabilize crop yield and to increase cropping intensity. The quantity of water to be applied as irrigation should be just enough to take care of the critical period and to cover as large an area as possible. Since the amount of water is sufficient for only one or two irrigations, it is essential to decide the time of irrigation. Supplemental irrigation can be scheduled by two approaches viz. soil moisture depletion approach and critical stage approach. Irrespective of the stage of crop, irrigation is scheduled when soil moisture approaches permanent wilting point to save the crop. This is called life-saving irrigation. Scheduling irrigation at critical stage of crop gives more return per unit area of water applied.

For crops like sorghum, flowering seems to be most critical stage. In tobacco, an irrigation three weeks before topping has given maximum benefit. The yield increase due to life saving irrigation varied depending upon the crop and the region.

Water stored in farm ponds is generally a small quantity due to limited catchment area and management practices adopted. The usual amount of water given per irrigation is 5 cm under irrigated conditions. As the amount available is less under dryland conditions and large areas are to be irrigated, a minimal irrigation method is adopted. Irrigation water is applied through a furrow 10 cm wide and 20 cm deep. The amount of water applied in this method ranges from 10 to 20 mm per irrigation to 20 mm per irrigation.

In situ moisture conservation

The second approach is to utilize the water where it falls by means of appropriate land treatments and in situ (i.e. at the same place) moisture conservation practices.

Some of these include off season tillage, mulching, dead furrows key line cultivation, compartmental bunding, graded border strips, and interplot water harvesting. Off-season tillage helps the rain water to enter into the soil profile more effectively and in addition, helps in weed control. Such a practice in alfisols (i.e. highly leached red brown soils) of the Telengana region in the state of Andhra Pradesh increased the sorghum yield by 43 per cent. Off-season tillage, however, is not suggested for aridisols (i.e. sandy soils of the desert regions) as this would accelerate wind erosion. Deep ploughing (i.e. > 220 cm), once every two or three years, also helps to increase crop yield in soils with a hard subsoil below the plough layer.

Mulching is a well-known practice for minimizing the movement of moisture from the soil into the atmosphere. Poor infiltration of rain water is a problem in vertisols (i.e. heavy black soils), which often leads to water stagnation. Vertical mulching (e.g. placing sorghum stubbles in trenches 40 cm deep, 15 cm wide, and protruding 10 cm above the ground level) has been found to enhance available soil moisture by 4-5 cm and to increase the grain yield of sorghum by 40-50 per cent in vertisols at several locations in India. The making of "idle furrows", also called "dead furrows" at each 3.6 m interval helps in improvising water retention. Such treatment at Anatapur in South India was found to increase the grain yield of groundnut by about 10 per cent.

Ridge-and-furrow *in situ* water harvesting is yet another promising system for the inceptisols (i.e. soils with poor horizon development as a in alluvial soils) of sub-humid areas. Maize is planted on the ridges and rice in furrows, thus meeting the varying water requirements of crops. The water requirement of rice is met by the water that is harvested into the furrows. Yield increases of about 18-21 per cent for both crops can be obtained by this system, as compared to flat sowing.

The inter row water harvesting system is a practice suitable for light textured soils found in the arid region. Significant yield advantages have been recorded with the pearl millet crop by adopting this practice, in combination with the recycling of the water for supplemental irrigation.

Runoff harvesting and recycling

Rainwater can be utilized by collecting in small dug out ponds excess runoff during high intensity rainfalls and then recycling it for supplemental irrigation at critical stages of crop growth. This water also can be used for pre-sowing irrigation of a rabi crop in potential double-cropping regions. A rabi crop is a crop grown during the post-rainy season (i.e. October-March) on stored soil moisture. Extensive studies have been undertaken on these aspects both in arid and semiarid areas. Experiments on catchment size and slope, size of the pond, effective sealants, water lifting devices, and choice of crops have been conducted both in alfisol and vertisol regions.

Existing water harvesting system suffers from the fact that the donor catchment (i.e. actual area that is contributing the runoff) does not benefit from the runoff water. In a modified system with small dugout ponds, the water can be used for the donor catchment. Studies have shown that appropriate pond size varies with rainfall, ranging from 200 cu m to 3000 cu m. Seepage losses pose a major problem in these dugouts. Though natural silting minimizes seepage losses in large tanks, it causes a reduction in storage capacity in small dugouts. Among the sealants evaluated, cement plus soil in a 1:8 ratio was found to be reasonably efficient and cost effective. But an ideal sealant has not been found yet that is both efficacious and cost effective.

Among the water application methods tried, an irrigating alternative furrow system was found to be more effective than use of the furrow method and flooding. Despite the clear demonstration of benefits of in situ moisture conservation and runoff recycling in both the experimental plots and the operational research projects, these practices have not yet been applied on a wide scale by the farmers. There are many constraints, both financial and technological, but the most critical constraints is the fact that farm holdings in India are very small and fragmented. The average size of an operating farm is about 1.8 hectares, which limits the application of conservation methods. Therefore, it is now realized that water conservation development projects should be undertaken on a watershed basis. The impact of water resource development can be better seen when improved practices are adopted in a watershed as a whole (i.e. a unit of land and a drainage area contributing runoff water to a common collecting point).

Efficient utilization of water through soil and crop management practices

There are quite a few options in the area of soil and moisture conservation. In most of the situations, contour cultivation and contour sowing will serve as good measures for soil and water conservation. In other situations where slope is more than 1.5 per cent, mechanical measures like contour bunding in low rainfall areas will be appropriate. In deep black soils where water stagnation could be a problem, ridge and furrow, or a broad bed and furrow (BBF) land configurations will be desirable.In soils having a hard pan or compact layer, deep ploughing once in 3 to 4 years may have to be resorted to for ensuring better infiltration of water and root growth.

Incorporation of crop residues, use of surface mulches, vertical mulching, impounding of runoff water in a dug out pond etc., are other soil and moisture conservation measures. In arid areas, setting up of shelter belts / wind breaks is an effective measure of soil conservation.

It is now established that contour farming parallel to graded bunds / key lines helps in conserving by both soil and water. For in situ moisture conservation, furrows are made with a plough after every row, 3-4 weeks after sowing - an operation which is incidental to interculture done in dryland crops. Experience gained in the Operational Research Project at Hyderabad has revealed an additional advantage of about 15 per cent in the yield of sorghum and castor, on account of furrowing as a part of interculture operation done on contour. The practice is easily acceptable by small farmers as it does not involve any additional cost.

As a first step, simple agronomical practices like off-season tillage across the slope, key line cultivation, flat sowing and ridging later, strip cropping, intercropping system have to by adopted. These practices can be easily adopted without much additional cost. Later, mechanical structures may be introduced after gaining full confidence of the farmers. The emphasis should be on in situ moisture conservation measures. Other soil management practices that have been developed for specific situations include compaction of highly permeable soils of low retentively, chiseling soils having layers of high mechanical impedance at shallow depths, raised/sunken bed technology for slowly permeable flat sols for regions receiving more than 1000 mm seasonal rainfall, etc.

In the past, mechanical structures were considered to be the panacea for all soil erosion problems. We never thought of vegetative barriers which could be as efficacious as mechanical structures. Where the slope is steep and the gradient is high, there is no substitute for mechanical structures. There are situations where the sol erosion problems are not that severe and the gradient in the waterway is not that high, vegetative barriers could be established with advantage. At times, we may have to combine, the mechanical drop structures with vegetative barriers. The whole approach will depend on the situation.

Moisture conservation in crop production

Soil moisture conservation management associated with timely operations plays a vital role in dryland areas in water is not limited in quantity only but available within a specified period. Hence there is need to conserve the moisture to keep moisture supply through out the crop life. Under dryland conditions, soil moisture conservation and management associated with timely operations play a very vital role in crop production since soil moisture is not only limited in quantity but it is also available within specified but uncertain periods.

Method of moisture conservation in dryland areas. Rainfall water can be conserved in following ways :

A) By storing more of rainfall in root zone.

1. Tillage,
2. Contour farming of bunding,
3. Vertical mulching,
4. Subsurface barrier,
5. Addition of pond sediments of organic matter
6. Addition of gypsum

B) By checking loss of water through evaptripiration

1. Mulching,
2. Weed control
3. Use of antitranspirants,
4. Crop thinning

C) By water harvesting

1. Water harvesting *in situ.*
   i) Inter row water harvesting
   ii) Inter plot water harvesting
2. Water harvesting for recycling

A) Increasing water storage in root zone.

1. Tillage: Tillage opens the soil make it friable and thus it helps in reducing runoff and increases infiltration of rainwater in this way it conserves the maximum amount of rainwater . But it is to be taken into account that whether we should go for deep tillage or shallow. Deep tillage is better than shallow tillage. Because it conserve more moisture and helps in control of weeds.
2. Contour farming bunding: On slopy lands, when tillage operation like ploughing, seed bed preparation, seeding, interculture operations are performed across the slope as nearly on contours as possible. It creates series of miniature barriers to water and thus reduces runoff as well as soil loss.
3. Vertical mulching: Vertical muching is a recent innovation which is found to be helpful in situ conservation of moisture for increased crop yields. This has been specially found to be suitable to black soils of Decan plateau whose intake rate are very low. It consist of jowar stubbles kep in trenches of 40 cm deep, 15 cm wide protruding 10 cm above ground level. Such trenches spaced at 4-5 meters increased crop yields by 400-500% in drought years and 40-50% in normal years ever control.
4. Sub-surface barrier: The practice is suitable for sandy soils where water is lossed through percolation. The placement of subsurface barrier of asphalt (2 mm thick at 60 cm depth) resulted in increased up take of N by bajra plants. These were associated with higher WUE and yield of crop.
5. Additional of pond sediments of organic matter
6. Application of gypsum

   In alkali-soils application of gypsum resulted in increased infiltration rates.

B) By checking evapotranspiration losses of water

1. Mulching

With respect to dryland mulch is any material placed at soil surface with a view to conserve the soil water. These may be various types such as - Straw or residue mulch, soil mulch, plastic mulch, and chemical mulch.

a) Straw mulching can help in conserving soil moisture in following ways :

   i) By checking loss of water from soil through process of evaporation because it reduces heating of soil from radiation and reducing wind speeds near soils surface.

   ii) By reducing runoff and increasing infiltration of water.

   iii) By checking weed growth.

b) Soil mulching: Surface mulch of dry soil 5-8 cm deep by obstructing the rise of water to the surface through capillary action effectively reduced loss of water as compared with effectively reduced loss of water as compared with a soil having an undisturbed surface. Following kinds experiment and conclusion, a soil mulch for moisture conservation became standard procedure in dry regions and dicta such as two cultivation can repoace one irrigation, became very popular.

c) Chemical mulching: It has been found that Hexaderunol, a long chain alcohol mixed with surface 1/4 inch of the soil reduced evaporation by 43%, this material which is resistant to microbial activity remained effective for more than a year. The surface layer of treated soil dried out more rapidly than that of untreated soil, creating a diffusional barrier to evaporation.

2. Weed control

Since time immemorial, controlling weeds has been known to be one of the most effective means of increasing the amount of water available to the crops and therefore, of increasing WUE. Weed frequently transpires greater amount of water per unit of dry matter produced than do the plants with which they grown in association. It has been estimated the amount of water saved by eliminating weeds in a maize field is equivalent to providing an entire irrigation at a time of maximum need.

3. Use of antitranspirants

An antitranspirant is any material applied on transplanting plant surface with the aim of reducing water loss from the plant antitranspirants can reduce transpiration in following ways.

a) By increasing leaf resistance to water vapour loss.

b) By reducing the net energy uptake by the leaves, through increasing leaf reflectance.

c) Reducing top growth by growth retardents.

Antitranspirant are of four types

i) Stomata closure e.g. PMA (Pheny) Murcuric scetate)

ii) Film forming : e.g. Plastic or waxy material.

iii) Light reflecting type - e.g. Kaolin.

iv) Growth retardant. e.g. CCC (2 chlorethl trimethyl ammonium chloride)

4. Crop thinning

Optimum plant population is one of the important practices for getting good yield under limited moisture conditions. Crop thinning may be a mid season correction for dryland agri. Thinning of plant population by removing very third row was found to be advantageous to mitigate the moisture stress.

(C) Water harvesting for reuse

Water harvesting is a technology of runoff farming, which aims at increase runoff and decreasing infiltration rate of rain water. It is not only suitable infiltration rate of rain water. It is not only suitable in arid and semi arid regions but also in sub humid regions where annual rainfall is sufficiently high but along dung spells do occur during crop growth.

Water harvesting is of two types :

i) Water harvesting in situ.

a) Inter row water harvesting.

b) Inter plot water harvesting.

ii) Water harvesting for recycling

Reducing water losses through mulches and use of antitranspirants

Dryland agriculture is practiced in most of the arid and semi arid areas. The main problem of these regions is low rainfall and meagre availability of irrigation water. A major portion of rainfall is lost in unproductive losses like evaporation, deep drainage, and seepage, there-by leaving a very small amount of water for Crop use. The crop-water use can be enhanced and water use efficiency (WUE) increased by controlling unproductive losses of water. Water use efficiency is the yield of marketable crop produced per unit of water used in evapotranspiration. It can be expressed as WUE = Y /ET, where Y = yield of marketable crop and ET = evapotranspiration or seasonal water use. The term evapotranspiration (ET) and water use (WU) are generally considered identical. These are, however, different in the sense that the former term refers to loss of water in both evaporation from the bare soil and transpiration from the crop, while 'the latter is the beneficial utilization of water for producing dry matter and seed. Therefore, for achieving higher water use efficiency the evapotranspiration losses need to be minimized and the water efficiently utilized for producing higher yields. This is particularly important in dry areas where water is both a scarce and precious commodity. This warrants a discussion on water losses and their control for achieving higher water use efficiency.

Evaporation

Evaporation is a process of losing water from a moist or wet surface to the atmosphere due to vapour pressure differences. From the moist soil surface the water is lost at about the same rate as from a free water surface having the same exposure and temperature. Harrold *et al.* [1959] reported that during the growing period of an annual crop about half the evaporated water comes from the soil. It, however, depends upon the type of soil, climatic conditions, the crop and its stage of growth. In arid regions, the loss of water due to evaporation from the soil is relatively low, the main component of lossing plant transpiration. It is because the soil dries quickly in such areas and acts as a mulch.

Evaporation process

Three conditions are necessary for evaporation to take place. First, there must be a continuous supply of heat to meet the latent heat requirement (590 cal/g of water evaporated at 15°C). This can come from the evaporating body or from outside as radiated or advective energy. Second, the vapour pressure in the atmosphere over the evaporating body must remain lower than the vapour pressure gradient between the body and the atmosphere and the vapour pres-sure must be transported away by diffusion or convection or both. These two conditions are influenced by meteorological factors such as temperature, humidity, wind velocity and radiations etc. The third condition relates to the supply of water through the body to the evaporating surface. This condition depends upon the amount, energy and the conducting properties of the evaporating body. Therefore, the evapora-tion from a soil surface or drying of soil depends upon climatic factors as well as the conducting properties of soil from lower moist layers to the evaporating site. The evaporation or drying of the soil takes place in three stages. The stage one takes place when the soil is wet and is controlled by external meteorological factors and soil surface conditions. Stage second is governed by the rate of supply of water to the evaporating surface. In stage third soil surface becomes extremely dry and the movement of water is in vapour phase. This stage persists for a longer period. The rate of loss of water in this stage is extremely low. In arid areas the first stage persists for a very short time but the second stage prolongs depending upon the soil moisture conditions and its transmission properties.

Evaporation control

Evaporation from the soil surface can be controlled by (i) reducing external evaporativity, (ii) reducing energy supply to the evaporating site, (iii) Decreasing the ,Conductivity / diffusivity of the, soil, and (iv) reducing the potential or the, force deriving water upwards through the 'profile. Based on these principles various measures have been devised and are discussed as under.

Shelterbelts/Shelter barriers: In arid and semi arid areas wind is one of the most important factors responsible for increased evaporation and loss of water from soil. Hot dry winds quickly dry the soil surface and create vapour pressure gradient for continuous vaporization. Shelter-belts or shelter barriers are one or multi rows of trees, shrubs and other plants planted in the field or field boundaries across the wind direction for reducing external evaporativity. The importance and utility or shelter belts in controlling wind erosion, reducing evaporative demand and protecting crops from hot desiccating winds particularly in arid and semi arid areas has been widely recognised by many workers in the world. In India. It has been observed that vegetative barriers as shelter-belts placed in Lile path of wind reduces its velocity near the ground by exerting a drag on the wind and deflecting the wind stream. The effectiveness of shelterbelts in reducing wind velocity depends on many factors such as wind velocity itself, direction, shape, width, tree height and density etc.

Surface mulching

Mulching is a practice of covering the surface of soil with plastics, organic materials like straw, grass, and stone etc. reduce evaporation, to keep down weeds and also to moderate wide fluctuations in diurnal soil. temperatures. It controls external evaporativity and also reduces energy supply to the evaporating site by cutting off solar radiation falling on the ground. Its main function is limited to controlling first stage of drying which helps in improved moisture status, reduced soil temperature, besides checking seedling mortality and improving crop stand. Besides the above, mulching helps in increasing downward movement of water. Its storage deep in the profile escapes evaporation due to reduction in thermal gradients and exchange of vapours. The effectiveness of mulches in conserving moisture has generally been found to be higher under more frequency of rainfall, drought conditions and also during early periods of plant growth when canopy cover remains scanty.

Tillage

Ti1lage is a practice of loosening soil, killing weeds and increasing porosity with the help of some tool or implement. These attributes, in turn, affect soil-water-relationships, aeration status, thermal characteristics and the mechanical impedance to the root penetration. Though tillage increases moisture loss from the soil surface by increasing exposed area and the free exchange of vapours, the dry layer' thus created acts as a mulch and restricts the upward movement of water to the evaporating site by reducing diffusivity gradients. Though the water Loss from the surface is higher, the cumulative loss from soil profile is reduced with this practice. This is particularly observed in medium sandy loam type of soils. In arid, sols the beneficial effects of tillage are specially derived from increased water recharge, reduced crop-weed competition, better root growth and proliferation, improved soil thermal conditions, leading increased water use and water use efficiency.

Transpiration and its control

Transpiration is a process of water loss from The plant surface about 99% of water absorbed by the roots is lost to the atmosphere as transpiration. The loss is mainly from the stomatal pores present on the plant leaves. As the stomata are the focal points on the plant leaf surface where the vapour pressure gradient between the leaf, surface and the atmosphere is the steepest, therefore, these are important sites for the application of antitranspirants. Antitranspirants can be defined as the materials which decrease water loss from the plant leaves by reducing the size or number of stomatal openings, decreasing there by, the rate of diffusion of water vapours (Fuehring, 1973). Two most important points in the use of antitranspirants are: (i) the application of antitranspirants should restrict the water loss from the leaf surface without reducing photosynthesis, as carbondioxide diffuses through stomata and is necessary for photosynthesis, (ii) transpiration causes cooling of the leaf surface and the use of antitranspirants should not completely stop transpiration and thus raise the leaf temperature.

Antitranspirants can be effective in two ways. viz., (i) through films that coat the leaf surface, and (ii) chemicals that close the stomata. Synthetic films that are sprayed on the leaf surface should ideally be resistant to water and completely permeable to carbondioxide. This is, however, very difficult to achieve. Antitranspirants such as phenylmercuric acetate and certain alkey1 succinic acids act by inhibiting stomatal opening. Film forming antitranspirants (waxy or plastic emulsions) produce an external physical barrier to retard the escape of water vapour from plants. Silicones penetrate the leaf and may act directly on the wet cell walls and white reflecting materials (white wash or kaolinite) lower leaf temperature and reduce the vapour pressure gradient from leaf to atmosphere.

Role of Antitranspirants

The main role of antitranspirants is to conserve water which otherwise is lost through transpiration. Since efficient crop production depends upon timely availability of water, any water conservation practice which enhances this availability should be useful. This is particularly important in arid areas where droughts are frequent and crop growth suffers due to short availability of water. In irrigated fanning the interval between two irrigations can be increased by using antitranspirants and thus water can be saved. .

In many cases the primary purpose of using antitranspirants is not water conservation per se but improvement of plant performance by in-creasing plant water potential. Antitranspirants can be used to control temporary shocks to plants during transplanting. Fuehring (1973) reported the use of stomata inhibiting or film forming antitranspirants for sorghum grown with limited irrigation with 5-17% increase in yield. Under hot arid environment conditions, use of antitranspirants has been reported to check shrinkage and improve the size and quality of peach and olive fruits. Although antitranspirants have been found to reduce both transpiration and photosynthesis, They have been found quite useful in conserving water for making it available during stress periods for sustained growth and production especially in arid and semi arid areas.

Deep percolation and its control

Sandy soils occupy an, important place in -many arid and semi arid areas. These soils are single grained, have poor moisture retention and storage capacity, allow heavy leaching and per-colation losses and are thus infertile and unproductive. Soils of arid areas of western Rajasthan have been found to contain 58 to 71 % fine sand and 24-28% coarse sand. These soils retain about 6 to 9% moisture at 0.1 bar tension which comes to about 90 to 140 mm of water per metre of soil profile. During periods of high rainfall or high rainfall intensity about 40-50% rainfall is lost as deep percolation retaining there by, low amounts of water in the root zone for plant use. Though the deep percolation water in the strict sense is not a loss as this goes to recharge the ground water and is available to perennial vegetation, yet it is not available to the field crops or plants with shallow root system. Therefore, research efforts were made to intercept this water for utilization of the crops/orchard plants.

Sub-surface moisture barrier

Sub surface moisture barrier is an artificial layer of a material with less permeability placed at a suitable depth in profile for intercepting the free flow of water. The water thus retained over the surface of barrier is low tension water and is easily available to the plants during the periods of drought. The barrier concept originated from the observation that sandy soils with clay sub soil are world's most productive soils. Though sub surface barriers were found quite useful, their large scale use could not be achieved because of non availability of suitable machinery for placement, large scale soil disturbance and high cost involved. However, sub surface barrier of bentonite clay and pond sediments were effectively used for orchard crops and establishment of trees in the arid areas, An in-tegrated use of circular water harvesting technique with sub surface moisture barrier of pond sediments was found highly effective for the growth and yield of *Zizyphus*, (Ber) and establishment of *Acacia tortilis* (Babul) and *Prosopis cineraria* (Khejri), plants and reducing their mortality under rainfed arid conditions of western Rajasthan.

Soil compaction

Compaction is a process of bringing soil particles closer to attain higher microporosity and thereby increased water retention in soil. Though higher levels of compaction have been reported to produce high mechanical impedance to root penetration and thus adversely affect plant growth, low levels create favourable conditions by way of increased moisture and nutrient retention.

Agroforestry system

Agroforestry is yet another technique to control deep percolation. Different trees like *Prosopis cineraria*, *Acacia senegal*, *Acacia nilotica*, *Zizyphus rotundifolia*, *Zizyphus mauritiana* if planted in agricultural fields of the arid areas, utilize the deep percolation water efficiently, because of their deep root system. The system thus helps in meeting grain, fuelwood and fodder requirements beside protecting and preserving the environment.

Role of antitranspirant in dryland agriculture

Globally 2/3 of earth surface is occupied by water but paradoxically only 1 % ground water is vital for subsistence of life on earth. Crops grown exclusively on rainwater are classified on dry land crops. Out of total land on earth 1/4 of the surface (11 %) is cultivable. 280% is rainfed area, 22 % without soil microbes, 105 is waterlogged, 230% of soil is without nutrient and 6 % land is covered with ice. In Africa, only 6% total land is under dryland. In India, particularly is the country where cover of 70% land under rainfed farming and it is very essential to manage every drop of water received through rains. "Antitranspirant is defined as any material applied to transpiring plant surface to reducing water loss from the plant" The possibility of reducing plant transpiration by chemical without material reducing photosynthesis is of great practical importance in arid region. Where crop production is limited by lack of water, however the potential uses of antitranspirants are not restricted to water conservation but may also includes the maintains of more favourable water balances in plants. Particularly in situation at plant growth stages. The practical use of antitranspirants in field crops involves decreasing water loss from leaves by reducing the size/ no. of stomatal opening and thus decreasing the rate of diffusion of moisture vapour. However, photosynthesis in the leaf depends on supply of CO_2 diffusing into the stomatal cavity. The loss of water through plant needs to be controlled because only a small portion of water taken up by plant roots incorporated into the plant body. Thus the very low efficiency in water utilization by green plants often a tremendous challenge for research in methods of decreasing water loss from vegetations. Another methods that of CO_2 enrichment are of the air is intended tom directly increase the rate of photosynthesis. Generally when plant photosynthsing affectively and stomates are widely open. The mesophyll resistance may be considerable greater than the stomatal resistance. The total impact of such an increase in Rs on the evaporation rate will be greater than the anti- photosynthesis rate. The concept of antitranspirants is developed in 1960. Several chemicals are identified since in 1960 and are classified as (1) Film forming types, (2) Stomata closing types, (3) Radiation reflectant types and (4) Growth retardants types. These chemicals when sprayed or applied through soil can benefit the crop under adverse rainfall. Drought reduces the crop yield by 0-100% depending upon severity. Normally it is reduced by 20-60 %. Crop productivity is dependent upon how fast a plant can recover after a stress of 6-10m days. Prolonged drought can be drastically reducing yield to zero level. Drought during the critical phenological phase like flowering and grain development is highly detrimental. Transpiration is said to be unavoidable evil but it has several function to attend in crop cycle. For one tone of food, about 1000 tones of water are needed. Cereal and legumes need 400-500 liters of water/ kg of grain, while fruit and vegetables require 1000 liters of water/ kg of food produced.

According to Waggoner, the transpiration is represented by:

$$T = \frac{5H_2O}{r_s + r_a}$$

Where T – Transpiration

$5H_2O$ – Diference in vapore conc. between stomatal cavity and free air outside.

r_s – Stomatal resistance

r_a – Resistance of boundry air layer

Thus the transpiration can be decreased by:

- Decreasing $5H_2O$ in the above equation.
- Increasing stomatal resistance.
- Increasing boundary air layer resistance.

This increase/decrease can be done by use of antitranspirants.

Water requirement of some crops and critical stages of growth

Crop	Water required (mm)	Critical period (Days after sowing)
Sorghum	400-500	28-30
Bajra	250-300	50-55
Maize	400-420	30-45
Wheat	400-450	60-65
Groundnut	400-450	40-45
Sunflower	300-350	30-35
Safflower	250-300	50-60
Gram	250-300	60-65
Cotton	550-600	80-90
Pigeonpea	500-550	90-100

Types of antitranspirants

1. Stomata closing type

Most of the transpirants occurs through the stomata on the leaf surface. Some fungicide like Phenyl mercuric acetate (PMA) and herbicides like atrazine in low concentration serve as antitranspirants by inducing the stomatal is closing. These might reduce the photosynthesis also simultaneously. PMA was found to decrease transpiration to greater degree then the photosynthesis in a number of plants the results of field studied have not be particularly increasing. Glyceryl half ester of decenyl succenic acid (GLOSA). Paragoute, Aliazine cycoeyl are the examples. Stomatal opening is regulated by various sensors like, water, CO_2, light and hormones. The opening is strongly controlled by hydroactive mechanism while other sensors are hydropassive and is mediated through RWC of guard cell chloraoplast. The ATs curtail the water requirement but it lead to proportionate reduction in Ps.

- The chemical should be impermeable to water and permeable to CO_2.
- It should be non – Toxic to he plants and animals.
- It should have longer retention and greater specificity.

- It should be degradable with solar UV radiation.
- It should be relatively inexpensive and easily available.

Recently, methanol (10%) is shown to have AT property and reduces stomatal conductance and transpiration.

PMA and ABA are capable of reducing stomatal aparture. The exact machanism is not known but it is thought that they may alter the permeability of the guard cells. Thus they decrease transpiration. The problem with these are:

By reduction in transpiration leaf temp becomes high and this will tend to increase transpiration and respiration and thus may decrease WUE.

Decrease in photo synthesis rate.

These chemicals are toxic and we are interested in non-toxic material which have long lasting effects.

2. Film forming type

These antitranspirants when sprayed on plants provides a physical barrier on the stomata. These will escape the loss of water vapour from the leaves. But the problem with these is they hinder the exchange of CO_2 & O_2 which ultimately affect the photosynthesis rate.

3. Reflectant type

These are white material, which form a coating on the leaves and increase the leaf reflectance (albedo). By reflecting the radiation, they reduce the leaf temperature and vapor pressure gradient from leaf to atmosphere and thus reduce transpiration. Application of 5% kaolin spray reduces transpiration losses. Kaoline and Bordeaux mixture (white ash, $CuSo_4$) are important chemicals under reflecting type. The objective is to apply a reflective pigment which increase the albedo and thus decrease the net radiation load and leaf temperature. 400-700 nm is useful radiation. Albedo is reflected light as the % of incident light.

4. Growth retardants

These chemicals reduce shoot growth and increase root growth and thus enable the plants to resist drought. They may also induce stomatal closer. Cycocel is one such chemical useful for improving water status of the plant.

Effect of antitranspirants (ATS) in crops: Adequate available water during the period from tollering through grain filling in cereals in necessary for production of high yield in semi arid regions. ATS may be used to retard the rate of soil water depletion.

Physiological effects

Physiological effect are mostly centralised on photosynthesis and transpiration (TR) because of its on gas exchange. The ATS reduce the rate of TR by limiting the vapor exchange through the stomata. It has also found that adverse effect on photosynthesis.

Effects of yield and yield attributes

Out of two antitranspirants Kaoline and PMA, the Kaoline increase no. of panicle, 1000 grain of weight, spikes length, no. of grains/ spikes more than PMA. However, increase in yield attributes of raya due to application of both PMA and Kaoline. Bringle again reported significant increase in yield of spring wheat grown in greenhouse due to PMA.

The following precautions are required to be necessarily followed:

1. Spray after rainfall stops.
2. Ensure two sprays of AT at an interval of one week.
3. Surfactants (Teepol 0.1 % detergent powder, 1 g/I) should be mixed in spray for effective response.
4. Repeat spray of AT if rains follow after one hour of spray.

Conclusion

1. Yield levels can be optimized
2. Better crop growth is assured.
3. Normal size grain is obtained.
4. Seed quality is improved and produce can be used for seed purpose.
5. Crop life is saved with marginal crop productivity.
6. Irrigation can be reduced in post rainy season crops.
7. If applied through drip irrigation (Cetyl alcohol or hexadecanol, the irrigation frequency can be minimized.
8. Some yield are assured when no yield are expected out of crop sown during severe drought.
9. Drought severity can be monitored
10. Losses can be minimized with limited inputs.
11 Fast reducing soil moisture can be arrested for better growth of crops.
12. Very useful for farmers having minimum irrigation facilities.

List of various types of antitranspirants

S.No.	Name of chemical	Types/ class	Mode of application
1.	Adol-52 (cetyl alcohol)	Soil water binding	Soil
2.	OED (Octadecanol)	-do-	Soil
3.	S-800	plastic	Spray
4.	Alkanyl suyccinic acid	stomata closing	Spray
5.	Sodium-Hydroxydecanol sulphonate	-do-	Spray
6.	Monoethyl etherof decenyl succnic acid	Stomata closing	Spray
7.	8-Hydoxyquinoline sulphonate	Reflecting	Spray
8.	Kaoline	Stomata closing	Spray
9.	Abscissic acid (ABA)	-do-	Spray
10.	Cycosil	-do-	Spray
11.	Ascorbic acid	-do-	Spray
12.	Phenyl murcuric acetate (PMA)	Reflecting	Spray
13.	Calcium bicarbonate	-do-	Spray
14.	China clay	Stomata closing	Spray
15.	TIBA	Plastic	Spray
16.	Waxol (wax)	Plastic	Spray
17.	Mobileaf(Trade name)	Stomata closing	Spray
18.	Folicot (Trade name)	-do-	Spray
19.	Hico 110 R(Trade name)	Plastic	Spray
20.	Polyethyline vinyl acryline	Stomata closing	Spray
21.	Phosphon	Stomata closing	Spray
22.	Paclobutrazol	Stomata closing	Spray
23.	Methanol	Stomata closing	Spray
24.	2,4-D	Stomata closing	Spray

Water use efficiency

As expected, water use efficiency (WUE) is increased by use of antitranspirants, especially under moisture stress conditions. PMA increased WUE of wheat at I x 10-4 M but decreased it at higher concentrations (7.5 x 10-4 M). Giri et at. (1983) also noted increased relative leaf water content and WUE of wheat after using PMA or kaolin. Antitranspirant spray on potato plants during tuber enlargement reduces plant water use and Increases the yield of large tubers (Increased WUE). Increased relative leaf water content and Increased WUE of chillies sprayed with PMA or kaolin at low soil moisture contents have been recorded. Effects of film-forming and silicone antitranspirants were explored on four herbaceous plant species. Slight increase in WUE was noted following the treatment with a dim-forming antitranspirant.

Moisture conservation

Evapotranspiration of wheat plants sprayed with 1 x 10-4 M PMA at 40, 60, and 80 per cent soil moisture depletion was 44.92. 42.52 and 38.35 cm, respectively. Kaolin-sprayed wheat plants grown under dryland saved 36 cm water. Kaolin-sprayed barleys maintained more available moisture in the soil. The consumptive value and the soil moisture depletion pattern were markedly influenced. Kaolin used in conjugation with PMA reduced daily evapotranspiration significantly. The reduction lasted for two weeks under clear weather. Irrigations, one at crown root initiation and the second at (lowering, produced similar yields to irrigation at all four stages. Spraying at crown root initiation and flowering with irrigation at jointing and grain filling, gave significantly higher yields than irrigation at crown root initiation and flowering with spraying at jointing and grain filling.

Variability in response to antitranspirants

All crop species do not respond similarly to antitranspirant treatments, even the varieties differ in their response. Season, region, stage of crop growth, and management practices also influence the effect of antitranspirants, both qualitatively and quantitatively. Twelve crop species Including eggplant, melon, cucumber, and grapevine were examined. The efficacy of the antitranspirant (hydrasyl) varied widely with plant species and depended on both physiological and morphological factors. Triadimetron reduced transpiration and Increased yield of soybeans and peas but not of wheat. Transpiration was reduced five to six days after treatment, the effect being greater In oats than In wheat. Maize and barley plants were sprayed with 2 per cent emulsions of paraffin oil or textol at three- to four-leaf stage. Paraffin oil gave lower water losses in maize while textol gave lower losses in barley.

Effects on growth and yield

Although the rate of photosynthesis is slightly reduced after the use of antitranspirants, the plants water economy is improved and wilting is avoided. Plants continue to grow at a lower rate than well-irrigated plants, but at a higher rate than un sprayed plants. Thus, the growth and yield of antitranspirant-sprayed plants are improved under rainfed dryland conditions. Kaolin spray on wheat increased grain yield due to increased tiller survival, longer ears and greater spikelet fertility. The number of shrivelled grains per ear was decreased. A combination of mulching and transpiration suppression brought about a saving of one irrigation on clay loan soil and at least two irrigations on a loamy sand soil.

Role of mulches in dryland agriculture

Mulch is any material at the soil surface that was grown and maintained in place; any material grown, but modified before placement and any material processed, manufactured or transported before placement. The "Materials include crop residues, leaves, dippings, bark, manure, paper plastic films, petroleum products, gravel and coal. The practice of applying mulches to soil is possibly as old as agriculture itself. Ancient Romans placed stones on the soil surface to conserve water and Chinese used pebbles from streambeds. The current trend under mechanized agriculture is to use crop residues as mulches on cultivated areas, usually in conservation tillage systems. An alternative is to use transported and manufactured materials as mulches for some high value crops. Mulches are used for various reasons but water conservation and erosion control are undoubtedly the most important for agriculture in dryland (semi-arid and arid) regions. While the effectiveness of mulches for water conservation is highly variable, mulches when properly managed definitely aid wind and water erosion control. Other reasons for mulching include soil temperature moderation, soil nutrient effects, soil salinity control, soil structure improvement, crop quality control and weed control. Considerable research on the use of mulches has been conducted in these aspects in dryland areas.

Effects of mulches on some soil properties and conditions

Mulches affect numerous soil properties and conditions either directly or indirectly. Among these affected are soil water content through run-off control, increased infiltration, decreased evaporation, and weed control; soil temperature through radiation shielding, heat conduction and trapping and evaporative cooling, soil nutrients through organic matter additions, differential nitrification, and mineral solubility, soil structure, soil biological regime through organic matter additions, microbial and soil fauna populations, and plant root distributions, soil erodibility and soil salinity through leaching and evaporation control. Probably the greatest importance, however, for agriculture in dry regions are soil water, temperature, structure and salinity. The effects of mulches, on these properties and conditions are briefed below:

Soil water

By their very nature, semiarid and arid regions generally receive. inadequate precipitation for good crop production. Furthermore, much of the precipitation that is received is lost by runoff and/or evaporation. Also, some water that enters the soil may evaporate before a crop is planted. Although, precipitation is limited and erratic and the evaporation potential is high, good yields can be obtained in semiarid and arid regions if most of the precipitation that occurs is effectively conserved and subsequently used for crop production. Numerous studies have been conducted to determine the influence of mulches on soil water storage, content, and evaporation. Many different materials have been used, but besides crop residues, one of the more widely researched has been plastic film (included are polyethylene, polyvinyl chloride and similar films.

Although, plastic films generally resulted in higher water content and reduced evaporation as compared with bare soil, they are relatively expensive and difficult to manage, especially under large-scale field conditions for low value crops. Consequently, less expensive and more readily applicable materials have been sought. Petroleum and asphalt sprays and resins have received considerable attention. These materials affected soil water in a manner similar to plastic films. Their effectiveness depended on the amount of surface covered. Most research with mulches has involved crop residues and other plant waste products (e.g., straw, stover, leaves, corn cobs, sawdust, and woodchips). These materials are cheap and often readily available, and They permit water to enter the soil readily. When maintained at adequate levels, these materials increase infiltration and soil water contents and reduce evaporation.

Soil temperature

When surface coverage was adequate, mulches generally resulted in greater soil water contents and lower evaporation than bare soil; however, the effects of mulches on soil temperatures were highly variable. Colour of plastic mulches greatly affected soil temperature. White or reflective plastic decreased temperatures Adams (1962) stated that clear plastic result in higher temperatures because the soil directly absorbs most of the energy from incoming solar radiation. The effectiveness of clear plastic for increasing soil temperatures increased with aging of the plastic. With black plastic, the soil received only a portion of the incoming energy absorbed by the film. Apparently, degree of contact between soil and plastic influences the extent of temperature increase under a black plastic. White and reflective plastics resulted in cooler temperatures because they did not absorb the incoming radiation.

Petroleum spray and resin mulches consistently resulted in higher temperatures than bare soil. The maximum difference usually occurred during the hottest part of the day. At night, the mulches had slight or no effect on soil temperatures. These materials (petroleum, asphalt, manure) absorbed the incoming solar radiation and, because of their close contact with the soil, readily conducted heat to the soil. Hence, the consistently higher temperatures with these mulches as compared with the variable influence of black plastic.

Mulches of plant materials (e.g., straw, stover, leaves) reduced soil temperatures. The combined effects of radiation in tercepti on and evaporative cooling, were responsible for the lower temperatures under this type of mulch. The 0-9 t/ha range reduced soil temperature at a 10 cm depth by 0.18 °C.

A group of miscellaneous mulching materials variously, and somewhat unpredictably, influenced soil temperatures. Black granular materials (e.g., coke, bitumen) increased temperatures, whereas lighter- coloured gravel reduced temperatures as compared with bare soil. The dark materials absorbed more radiation than the lighter-coloured gravel; hence, the higher temperatures with the black materials.

Soil structure

In dry regions, precipitation frequently occurs in high-intensity storms. When falling raindrops strike bare soil, soil particles are dispersed and surface sealing may occur, thus, reducing infiltration. Consequently, water that could be conserved for plant use is lost by runoff. The dispersed soil at the surface often forms a hard crust when dry, which may adversely affect seed germination. Seedling emergence, plant growth, and soil aeration. Beneficial effects of surface mulches on soil structure result primarily from mulches absorbing the energy of falling raindrops, thus reducing soil dispersion and surface sealing. Infiltration rates generally are maintained, and subsequent crusting is reduced. Other benefits possibly result from greater microbe and fauna activity in soil, greater root proliferation, and the cushioning effect, which reduces compaction due to tractor, implement, or animal traffic.

The degree of protection provided at the soil surface is related to surface coverage provided by the mulch. Plastic films prevent raindrops from striking the soil and, thus prevent dispersion. However, unless special provisions are made, they also prevent water infiltration. Less compaction also was ascribed to a plastic mulch.

Petroleum, bitumen, coke, and desert pavement mulches reduced surface crusting and soil density. The materials, except possibly petroleum, undoubtedly also resulted in greater water infiltration. Although effective for reducing crusting, these materials may not provide the necessary binding substances for improved soil aggregation. However, ii the surface remains adequately covered, stable surface aggregates are less important than where the soil is partly or completely exposed.

Crop residues and other similar mulches, when present or applied at adequate levels, have maintained high infiltration rates. The high infiltration rates resulted from less surface sealing due to the protection against falling raindrops afforded by the mulches. Decomposition products of the mulches resulted in improved soil aggregation with respect to aggregate size, number and stability. Although, aggregation was improved, the aggregates were not sufficiently stable to withstand the impact of water drops without a protective cover.

Conservation tillage systems that maintain most crop residues on the surface have given variable resulted with respect to soil structure. Many authors reported larger water-stable aggregates and lower bulk densities as a result of stubble-mulch tillage when compared to clean tillage.

Soil salinity

Many soils in semi-arid and arid regions have a high salt content. Because some of the salts are readily soluble in water, they move with the water. The salts can be removed from the soil by leaching if precipitation is adequate. However, due to limited precipitation, they often move only to a limited depth and readily return to the surface with capillary flow as the soil water evaporates. Under such conditions, susceptible plants may be severely injured by the salts, especially at the germination and seedling stages. With irrigation, the salt problem may be aggravated by applying water having a high salt content and by applying insufficient water to move the salts well below the plant root zone. The management of saline soil has been widely investigated.

Because salts readily move with soil water, and practice that maintains infiltration rates and reduces subsequent evaporation should moderate the adverse effects due to soil salinity. Damage due to salts is most severe at germination and plant seedling stages; hence, any practice that reduces the salt content in the seed zone should be beneficial for plant establishment. While complete soil covers may be too expensive and impractical for extensive areas, strip mulches (e.g. plastic films, petroleum sprays) that maintain higher water content in and reduce evaporation from the seed zone should reduce the salinity hazard and aid plant establishment. Additional benefits should result if the mulches continue to reduce evaporation and if the crop yield and quality are improved due to salinity control.

A soil salinity problem associated with increased water conservation, as with stubble-mulch tillage under fallow conditions, is the saline-seep problem in some areas. The problem develops when water refills the soil profile and deeper layers at upslope positions. As the water flows down slope through the soil, it carries with if dissolved salts. The saline water seeps to the surface at downslope positions, adversely affecting crop production at the seep sites. The most effective solution to the saline-seep problem is to use as much of the current precipitation as possible for crop production before the water percolates beyond the root zone. Forage crops, especially perennials that have deep root systems, are effective for reducing the problem.

Effects of mulches on erosion control

Because of low residue production and dry soil surface conditions semi-arid and arid regions are highly susceptible to wind erosion. Water erosion also can be a problem because the precipitation frequently occurs as intense storms and the surface is inadequately protected by vegetation to effectively retard runoff. Therefore, mulches are important in semi-arid and arid regions for reducing erosion. The ease by which soil particles are moved by wind and water is related to particle size and to wind or water velocity. Although particles greater than 0.84 m in diameter are generally considered non-erodible by wind, water may cause erosion of particles of almost any size. Decomposition products of some mulching materials may Increase soil aggregation, thus reducing the susceptibility of soil to erosion. The chief value of mulches with regard to erosion control, however, is not their effect on soil structure but their moderating the wind and water forces at the soil surface. Additional benefits for water erosion control result from increased water infiltration rates and reduced surface runoff due to mulches.

A wind erosion equation has been developed which relates soil loss by wind to the (1) soil erodibility index measured in terms of soil clods greater than 084 mm in diameter; (2) climatic factor measured in terms of wind velocity and surface water content; (3) soil surface roughness; (4) unsheltered field width; and (5) vegetative cover. Based on equation, Woodruff (1972) has been shown than 1.1, 2.2 and 6.6 t/ha of standing wheat, flattened wheat, and flattened sorghum residues, respectively, were required to keep wind erosion to a tolerable 11.2 t of soil loss per hectare annually. The stubble-mulch farming system, widely used in North America, was developed primarily for wind erosion control The value of stubble- mulch tillage for erosion control is widely recognized. In general, erosion by wind decreased as the amounts of crop residues maintained on the surface were increased by stubble-mulch tillage. Wind erosion was even more effectively controlled where higher amounts of residue were maintained on the surface by chemical fallow or no tillage practices.

Standing residues were more effective for wind erosion control than flattened residues; but to shield against the erosive action of moving water, the mulch must oppose water flow at the soil water interface. Hence, a maximum amount of the material should be in contact with the soil surface. In general, erosion was reduced as the amount of mulch was increased. However, if the mulch was not in adequate contact with the soil, erosion could occur under the mulch. In recent years, no tillage systems have received considerable attention. With these systems, crop residues are maintained on the surface, weeds are controlled with chemicals, and subsequent crops are seeded with no more soil disturbance than necessary to place the seeds in the soil. Excellent control of erosion by wind and water has been obtained.

Effects of mulches on plants

The effects of mulches on plants are operative through their effects on soil water, temperature, structure, salinity, and erosion. Critical periods in the life cycle of a plant are germination, emergence and seedling establishment. For germination to occur, viable seeds must be placed in a favourable environment with respect to water supply, temperature, and aeration. After germination, the seedlings must emerge and become established. Due to their small size and tenderness, seedlings can easily be adversely affected by an unfavourable environment. Mulches can aid germination, emergence, and seedling growth & yield moderating or improving the soil and aerial environment to which the seeds and seedlings are subjected. Higher soil water contents and reduced evaporation were major reasons for improved germination, emergence, and seedling growth due to mulches. Straw, petroleum, gravel, stones, and plastic films increased germination and early growth; but the plastic films had to be slit or removed for continued growth. Higher soil temperatures due to mulches also improved germination, emergence and seedling growth. The higher temperatures, either alone or in combination with higher water contents, were also effective in promoting later plant growth and hastening plant maturity. Early seeding of sweet corn, made possible by using a clear plastic mulch, resulted in the corn being ready for harvest two weeks before corn seeded in bare soil.

Surface mulch, which maintained a lower, more constant soil temperature than other materials after irrigation were effective for improving growth of early potato *(Solanum tuberosum* L.). In other cases, lower temperatures due to mulches reduced germination and early plant growth. However later growth sometimes was better on mulched than on bare soil due to improved water conditions under the mulch. Many of the reports have indicated higher yields when crops were grown with mulches, rather than without mulches, with most mulching materials being effective for increasing yields. These ranged from about 50 to 300 per cent.

Yield response due to mulches, although primarily related to soil water and temperature, was affected also by better plant populations, reduced root rot. Undoubtedly, mulch effects on soil structure, nutrients, microbial activity, root distributions, and other factors also played an important role in the higher yields. Reduced crop yields due to mulches have been related to some specific conditions, generally soil temperature. Corn apparently is very sensitive to low temperatures under straw mulch early in the season when temperatures normally limit plant growth. The adverse effects of low temperatures were reflected in lower yields in more northern regions, but were largely overcome in the more southern regions.

Chapter 7

Management of Crops in Rainfed Areas

Soil and water conservation measures

Soil conservation is using and managing the land based on the capabilities of the land itself involving application of the best management practices leading to profitable crop production without land degradation.

Water erosion control

Water erosion occurs simultaneously in two steps : detachment of soil particles by falling raindrops and transportation of detached soil particles by flowing water. Therefore, water erosion can be minimized by preventing the detachment of soil particles and their transportation.

Principles of water erosion control are

- Maintenance of soil infiltration capacity
- Soil protection from rainfall
- Control of surface runoff, and
- Safe disposal of surface runoff.

For a sound soil conservation programme, every piece of land must be used in accordance with land capability classification. Measures of water erosion control can be broadly grouped into three: agronomic, mechanical and forestry measures.

Agronomic measures

In soil and water conservation programme, agronomic measures (conservation agronomy) have to be considered in coordination with others for their effectiveness. These measures are effective on gentle slopes up to 9 per cent. Reduction in runoff is achieved by choice of crops, land preparation, contour cultivation, stripcropping, mulching, application of manures and fertilizers and appropriate cropping systems.

Choice of crops: Row crops (erosion permitting crops) such as sorghum, maize, pearlmillet, etc. are not as effective as soil conserving crops (erosion resistant crops) such as cowpea, groundnut, green gram, black gram, etc. Generally, legumes (smothering crops) provide better cover and protection to soil by way of minimising the impact of raindrop and acting as obstruction to runoff.

Land preparation: Land preparation including post harvest tillage influence intake rate of water, obstruction to surface flow and consequently the rate of erosion. Deep ploughing or chiselling has been found effective in reducing erosion. Rough cloddy surface is also effective in controlling erosion.

Contour cultivation: All the cultural practices such as ploughing, sowing, intercultivation, etc. across the slope reduce the soil and water loss. By ploughing and sowing across the slope, each ridge and plough furrow and each row of the crop act as obstruction to the runoff and provide more time for water to enter into the soil leading to reduced soil and water loss.

Strip cropping: It involves growing of few rows of erosion resisting crops and erosion permitting crops in alternate strips on contour with the objective of breaking long strips to prevent soil loss and runoff. Erosion resisting crops reduce the transporting and eroding power of water by obstructing runoff and filtering the sediment from the runoff to retain in the field.

Mulching: It reduces soil loss considerably by protecting the soil from direct impact of raindrop and reducing the sediment carried with runoff. A minimum plant residue ground cover of 30 per cent is necessary to keep runoff and soil loss within acceptable limits. As most of the plant remains are used as cattle fodder, availability of organic residues for mulching is the major limitation under conditions of scarce fodder.

Organic manures and fertilizers: Organic manures, besides supplying nutrients, improve soil physical conditions. Fertilizers improve vegetative growth (canopy) which aids in erosion control. Improvement in soil structure improves rate of infiltration leading to reduced runoff.

Cropping systems: Monocropping of erosion permitting crops accelerate soil and water loss year after year. Intercropping erosion permitting and erosion resistant crops or their rotation have been found effective for soil and water conservation. As. the legumes (cowpea, greengram, horsegram, blackgram) are effective for soil conservation due to their smothering effect, they should be sown in time to develop adequate canopy by the time of peak rate of runoff.

Cover crops such as greengram, blackgram, cowpea, soybean, sunnhemp, groundnut etc restore soil fertility besides reducing soil erosion. Such crops as intercrops with widely spaced crops can give continuous cover of ground protection against erosion.

Forestry measures

Perennial vegetation such as trees and grasses can be successfully used for economic utilisation of degraded lands besides reducing soil and water loss. Effectiveness of perennial vegetation in reducing the impact of raindrop on soil surface and minimising the velocity of runoff of surface flow has been well established.

A large number of tree species were introduced for evaluating their performance under different agroclimatic conditions of the country. Tree species found economical are

- *Pinus patula* (Pinus)
- *Pinus kesia*
- *Eucalyptus camaldulensis* (Eucalyptus)
- *Moliodora*
- *Leucaena latisilqua* (Subabul)
- *Acacia inearnsii* (Acacia)
- *Acacia nilotica*
- *Acacia tortilis*

The desirable characters of grass for soil and water conservation are perennial nature, drought resistance, rhizomniferous, good canopy, deep root system, prostrate habit and usefulness in cottage industry. Some of the useful grasses and legumes are

Grasses

- *Cenchrus ciliaris*
- *Chloris guyana*
- *Cynodon dactylon*
- *Dicanthitun annulatuin*
- *Heteropogon contortus*
- *Iseilenia laxtim*
- *Panicunt antidotale*
- *Andropogan haltei*
- *Panicum virgatuni*
- *Eragrostis curvula*

Legumes

- *Alylosia scarbaceoides*
- *Centrosenja pubescence*
- *Siylosanthus hamata*

Hard, medium and software treatments for rainfall scarcity areas

Soil moisture conservation measures coupled with water harvesting help to improve the moisture availability in soil profile and surface water availability for supplemental off season irrigation. The interventions through land management measures have greater role to play on transferring a part of surface water to ground water by recharge. Based on the nature and type of hydraulic barriers and their cost, the conservation measures in arable lands can be divided into following three categories.

1. Hardware treatments
2. Medium software measures
3. Software measures

1. Hardware measures

These are generally of permanent type, provides for improvement of relief, physiography and drainage feature of the area. These are erected with the major Government support with the purpose to check soil erosion, regulate over land flow and reduced peak flow. At times, they are imposed to completely divert the upland runoff from running into down stream fertile lands. Different hardware measures are explained below.

a) Water ways: Even in drylands, quite often high run off volumes are observed particularly in watershed located in sloppy areas. Such runoff water should, therefore be channalised through a few water ways, as far as possible some of the existing water ways may be developed. Water ways draining larger areas may be designed on hydrologic and hydraulic consideration. No doubt, at place sizable land strips may be found highly advantageous for the survival of many multipurpose useful trees when planted along the water ways. Depending upon land situation alignment of small size water ways may be adjusted with field boundaries; at times, such arrangement are specially useful for safe guarding crops against over stagnation of rain water in black soil areas. Such water course may also be safe generated against erosion by providing mechanical checks and also by raising vegetation which may be useful fodder for animals. In some situation water ways may also have a small section side bund with necessary openings at water entry points.

b) **Bunds:** These are low height earthen embarkments constructed across the slope in cultivated lands after deciding the location of water ways. The bunds function to intercept runoff, increase infiltration opportunity time and dispose excess rainfall safely.

i) Contour bunds are recommended in dry farming areas with light textured soils of slopes upto 6 per cent and where annual rainfall does not exceed 600 mm.

ii) Graded bunds are constructed in medium to high rainfall areas in poor permeable soil (vertisols) having 2 to 6 per cent slope. These are also quite suitable for the soils having crust formation tendency like red "chalka" soils of Telengana region of Andhra Pradesh.

c) **Terracing:** cultivated lands having land slope above 10 per cent particularly in Hilly areas should be put under bench terracing by converting the lands into series of platform. The width of bench terrace depends upon the land slope and the permissible depth of the cut. Bench terracing is very much effective in reducing soil erosion in hilly areas.

2. Medium software measures

Medium software are also provided particularly as interbused treatment, where field sizes are large and conventional bunds are constructed along field boundaries. Such treatments are usually of semi-permanent type and are adopted to minimize the velocity of over hand flow. They may need major initiative from cultivators in addition to some grants from the government side. Such measures many last 2 to 10 years, vegetation component and land configuration may also provide some direct returns on short term basis. However, these need to be modified at time to maintain effectives for erosion control and moisture conservation, and also to minimize risk of providing shelter to harmful pests and around these measures and in certain situations like taking up corrective measures to avoid to much spread of introduced vegetation and being counter productive.

a) **Small section / key line bunds:** A small section bund may be created cross the slope at half of the vertical bund spacing. Such bunds can be nearly 0.1 m^2 in section; many be renovated at an interval of 2 to 3 years.

b) **Strip leveling:** About 4 to 5 cm strip of land above the bund across the major land slope may be levelled for the purpose. Similarly one or more strips can be created at mid length of slope. Such strips can be erected by running blade harrows after ploughing the field with mould board harrows after ploughing the field with mould board or disc plough. Such mirror range leveling programme may be taken up after every 2 to 4 years.

c) **Live beds:** One or two live beds (2-3 m wide) can be created either on contour or on grade in the inter bund space. The vegetation on the beds may be according to liking of cultivators, it can be annual, perennial or a combination of both.

d) Vegetative (live barriers): One or two barriers of close growing grasses or legumes can be created along bunds as well as at the mid length of slope to filter the runoff water or slow down over land flow. Khus could be one of such vegetation. Several other priming grass species that serve as valuaba fodder for cattle are being explored as an alternative to khus grass. A miniature bund at lower side of barrier is recommended to help in the development of live barrier particularly in the initial stage.

3. Software treatments

A mention was made that hardware type land treatment all useful for safe runoff disposal and similarly, medium software are essentially to slow down the velocity of over land flow in cultivated fields. However, on several occasion these are found inadequate in attaining equitable moisture distribution for crop growth. In such cases, software treatment are taken up for ensuring uniform soil moisture for satisfactory crop performances. By and large, software treatments are tenaporary in nature, in that case these are required to be ronade or renovated every year. The entire cost of applying such treatments are to be met by the farmers. Because of favourable economics, a few of these treatments have gained wide acceptances in the recent years.

a) Contour farming**:** Contour farming is one of the easiest and most effective and low cost method of controlling erosion and conserving moisture. With contour farming, tillage operation are required / carried out along contours. It creates numerous ridge and furrous for harvesting sizable amount of runoff inside the soil.

b) Compartmental bunds**:** Compartmental bunds, converting the area into square / rectangular parcels – are useful for temporary impending of water for improving moisture status of the soil. These are made using bund farmers. In a medium, deep black soils, they are found advantageous in storing the rainfall received during the season in soil profile thereby augmenting the soil moisture for use by rabi crops. The size of the compartments may be fixed considering the size of the interbunded land.

c) Scoob formation**:** Making small depression on soil surface. This practice can reduce the runoff of water by 69 per cent and solid loss by 53%.

d) Tile ridges**:** Ridges and furrow system helps in moisture increment in soil profile and can reduce the runoff. In tile we can reduce runoff by 39 per cent and solid loss by 28%.

e) **Dead furrows:** Dead furrows are laid across the land slope in rolling lands to intercept the runoff. The spacing between dead ferrous between 2 to 5 metre or 4 to 7 crops rows. This system works will in alfisols. (Singh and Rao, 1998).

f) **Broad bed and furrous:** Broad bed and ferrous system implies shaping alternative bed and ferrous. This technique is especially suited to black soils. Where crops are sown on preformed bed. This system is made before the season and is maintained year after year. The planting is done on the bed. Generally the depth of each furrow is kept 0.15 metre and the inter furrow spacing is maintained at 1.5 metre.

g) **Tillage:** Tillage operations help in rain water infiltration. Off season tillage, in particularly has been found quite useful in most rainfed areas. Tillage operations make the soil receptive to rainfall. This practice is very useful in light soils often prone to crusting.

h) **Mulching:** Surface mulching protects soil against beating action of rain drop and also it increased water infiltration into the soil. Further it help in minimizing water evaporation from the soil surface. Sometimes dry soil mulch created simply by stirring the soil has been found effective for good performance.

i) **Deep ploughing during summer:** Deep ploughing during summer particularly in soils with slowly permeable sub soil layer has been found to increase moisture storage in the root zone and consequently the yield. It has been reported that deep ploughing to 30 cm where sub soil had slowly permeable layer at Ambala and Ludhiana increased the moisture storage in soil and enhanced wheal yield by 14%.

j) **Inter row water harvesting:** In low rainfall areas the water is collected in already opened furrows rather than every distributing the rainfall in order to increase moisture content for raising crops. This technique is also useful or used to drain excess water and to provide aeration to growing crops in high rainfall areas. Water stored in furrow is easily available to crops. Experiments at International Crop Research Institute for semi and tropics and under AICRP on dry land have indicated that inter row water harvesting is quite useful for raising yield level in semi-arid area.

k) **Alternative planting pattern:** In areas where plant often experience moisture stress, particularly during long dry spell, the techniques of planting 'narrow' ridge furrow method has been found beneficial in making good use of water or rain water. In this system kharif crops are planted on ridges and rabi crops planted in ferrous subsequently in the winter season. Ridges serve as micro-catchment for furrow where rainfall accumulates. Water collected in furrows meet the water need at crucial period of plant life and also ensures better aeration for crop planted on ridges during the kahrif seasons. After harvesting of kharif crops, rabi crops are planted in furrow where sufficient moisture have been stored for germination of rabi crops.

Inter plot water harvesting**:** In drought pone areas, it may not be only feasible but desirable to leave a part of land for water harvesting and a part for cropping. The first portion, which is relatively elevated, is used for conveying water is runoff plot and the other one which receive the runoff water (relatively law land) is called runoff plot. Experiment at Jodhpur has shown that the total production by cropping only 2/3rd of the field (leaving one third) by adopting runoff farming is the same as obtained from cropping the whole field.

Moisture conservation in crop production

Soil moisture conservation management associated with timely operations plays a vital role in dryland areas in water is not limited in quantity only but available within a specified period. Hence there is need to conserve the moisture to keep moisture supply throughout the crop life.Under dryland conditions, soil moisture conservation and management associated with timely operations play a very vital role in crop production since soil moisture is not only limited in quantity but it is also available within specified but uncertain periods.

Method of moisture conservation in dryland areas. Rainfall water can be conserved in following ways :

(A) By storing more of rainfall in root zone

1. Tillage
2. Contour farming of bunding
3. Vertical mulching
4. Subsurface barrier
5. Addition of pond sediments of organic matter
6. Addition of gypsum

(B) By checking loss of water through evapotranspiration

1. Mulching
2. Weed control
3. Use of antitranspirants
4. Crop thinning

(C) By water harvesting

1. Water harvesting in situ
   i) Inter row water harvesting
   ii) Inter plot water harvesting
2. Water harvesting for recycling

(A) Increasing water storage in root zone

1. Tillage: Tillage opens the soil make it friable and thus it helps in reducing runoff and increases infiltration of rainwater in this way it conserves the maximum amount of rainwater. But it is to be taken into account that whether we should go for deep tillage or shallow. Deep tillage is better than shallow tillage. Because it conserve more moisture and helps in control of weeds.
2. Contour farming bunding: On slopy lands, when tillage operation like ploughing, seed bed preparation, seeding, interculture operations are performed across the slope as nearly on contours as possible. It creates series of miniature barriers to water and thus reduces runoff as well as soil loss.
3. Vertical mulching: Vertical mulching is a recent innovation which is found to be helpful in situ conservation of moisture for increased crop yields. This has been specially found to be suitable to black soils of Deccan plateau whose intake rate are very low. It consist of jowar stubbles keep in trenches of 40 cm deep, 15 cm wide protruding 10 cm above ground level. Such trenches spaced at 4-5 meters increased crop yields by 400-500% in drought years and 40-50% in normal years ever control.
4. Sub-surface barrier: The practice is suitable for sandy soils where water is lossed through percolation. The placement of subsurface barrier of asphalt (2 mm thick at 60 cm depth) resulted in increased up take of N by bajra plants. These were associated with higher WUE and yield of crop.
5. Additional of pond sediments of organic matter
6. Application of gypsum

   In alkali - soils application of gypsum resulted in increased infiltration rates

(B) By checking evapotranspiration losses of water

1. Mulching: With respect to dryland mulch is any material placed at soil surface with a view to conserve the soil water. These may be various types such as - Straw or residue mulch, soil mulch, plastic mulch, and chemical mulch.

a) Straw mulching can help in conserving soil moisture in following ways -
   i) By checking loss of water from soil through process of evaporation because it reduces heating of soil from radiation and reducing wind speeds near soils surface.
   ii) By reducing runoff and increasing infiltration of water.
   iii) By checking weed growth.
b) Soil mulching: Surface mulch of dry soil 5-8 cm deep by obstructing the rise of water to the surface through capillary action effectively reduced loss of water as compared with effectively reduced loss of water as compared with a soil having an undisturbed surface. Following kinds experiment and conclusion, a soil mulch for moisture conservation became standard procedure in dry regions and dicta such as two cultivation can replace one irrigation, became very popular.
c) Chemical Mulching : It has been found that Hexaderunol, a long chain alcohol mixed with surface 1/4 inch of the soil reduced evaporation by 43%, this material which is resistant to microbial activity remained effective for more than a year. The surface layer of treated soil dried out more rapidly that of untreated soil, creating a diffusional barrier to evaporation.

2. Weed control

Since time immemorial, controlling weeds has been known to be one of the most effective means of increasing the amount of water available to the crops and therefore, of increasing WUE. Weed frequently transpires greater amount of water per unit of dry matter produced than do the plants with which they grown in association. It has been estimated the amount of water sowed by eliminating weeds in a maize field is equivalent to providing an entire irrigation at a time of maximum need.

3. Use of antitranspirants

An antitranspirant is any material applied on transplanting plant surface with the aim of reducing water loss from the plant antitranspirants can reduce transpiration in following ways.

a) By increasing leaf resistance to water vapour loss.
b) By reducing the net energy uptake by the leaves, through increasing leaf reflectance.
c) Reducing top growth by growth retardents.

Antitranspirant may be of four types.

i) Stomata closure e.g. PMA (Pheny) Murcuric scetate)

ii) Film farming: e.g. Plastic of waxy material.

iii) Light reflecting type - e.g. Kaolin.

iv) Growth retardant e.g. CCC (2 chlorethl trimethyl ammonium chloride)

4. Crop thinning

Optimum plant population is one of the important practices for getting good yield under limited moisture conditions. Crop thinning may be a mid season correction for dryland agriculture. Thinning of plant population by removing very third row was found to be advantageous to mitigate the moisture stress.

(C) Water harvesting for reuse

Water harvesting is a technology of runoff farming, which aims at increase runoff and decreasing infiltration rate of rain water. It is not only suitable infiltration rate of rain water. It is not only suitable in arid and semi arid regions but also in sub humid regions where annual rainfall is sufficiently high but along dung spells do occur during crop growth.

Water harvesting is of two types -

i) Water harvesting in situ.

a) Inter row water harvesting.

b) Inter plot water harvesting.

ii) Water harvesting for recycling

Critical growth stages to water stress for major field crops

Crop	Crop duration (days)	Critical stages for irrigation
Wheat (dwarf)	130-140	Crowing, flowering, jointing, milk and dough
Wheat (tall)	140-150	Tillering, heading and dough
Barley	125-135	Tillering, booting
Rice (dwarf)	130-140	Flowering, grain filling, tillering
Rice (Basmati)	140-150	Flowering, grain development
Summer maize	90-110	Tasseling and silking, and grain development
Winter maize	150-180	Tasseling, grain development
Pearlmillet	75-85	Heading
Sorghum	90-110	Flowering, seeding, grain filling
Cotton	150-180	Flowering and boll development

Chickpea	130-140	Preflowering and pod development
Pigeonpea	150-170	Flowering initiation, pod development
Moongbean	65-75	Flowering
Clusterbean	130-145	Branching and flowering
Mustard	140-150	Pre-flowering and pod development
Groundnut	110-135	Pegging, pod-setting and pod filling
Sunflower	100-120	Flowering, seed filling, late vegetative
Sugarcane	280-330	Tillering and grand growth

Water requirement (WR) of major dryland crops and fruit trees

Summer crops	WR(mm)	Winter crops	WR(mm)	Fruit crops	WR(mm)
Pearlmillet	250	Rocket plant	150	Gonda	560
Mungbean	225	Mustard	200	Ber	660
Clusterbean	275	Chickpea	250	Aonla	660
Cowpea	240	Lentil	285	Clusapple	860
Pigeonpea	450	Peas	300	Guava	1160
Maize	400	Barley	300	Pomegranate	1160
Sorghum	400	Wheat	350	Mango	1260

Strategies for Sustained Food Production in Rainfed Region

Technology developed for management of crops in rainfed areas in various research institutes have been summarized below:

1. Identification of viable rainfed technologies

A number of economically viable rainfed technologies have been developed over the years in the country to address the problems of food production in rainfed agriculture through CRIDA and its network centresfor the last three decades. These technologies have been evolved after refining them in farmers' field through Operational Research Projects, Institute Village Linkage Program (IVLP) and farm science centres. These include simple practices like off-season tillage in rainfed Alfisols and related soils for better moisture conservation 3 and weed control. Farmers in Operational Research Project (ORP) areas of Hyderabad adopted this practice in sorghum and castor and realized yield advantage by 40 percent over traditional practice. Lack of adequate draft power with many small farmers, however, is one of the major constraints to popularize this practice. Custom hiring of tractor is effective solution of farm mechanization on these lands.

2. Soil and rainwater conservation techniques

Efficient conservation of rainwater is the central issue in successful dryland farming. Extensive trials conducted by the soil conservation and dryland research centres have led to the identification of a number of inter-terrace land treatments besides contour and graded bunds. These techniques are location specific and the benefits from their adoption are highly variable depending on the rainfall intensity, slope and texture of the soil besides the prevailing crop/cropping system. Farmers have not widely adopted mechanical measures like contour bunds, graded bunds, grassing of waterways and construction of farm ponds without the government support due to financial constraints. However, studies at Hyderabad, Bangalore and Anantapur revealed that more than 80 percent farmers follow simple conservation measures like sowing across the slope, opening of dead furrows and key line cultivation. The yield improvement by adoption of soil and water conservation measures vary between 12 and 20 percent which are at times not convincing enough to farmers. However, cumulative effects are significantly visible at some locations. Since such measures help in long-term conservation of resources, these are implemented through the Government of India or the respective State Government sponsored watershed management programmes.

3. Timely planting of crops

Timely sowing and precision are essential for getting good plant stand, higher yield and optimum utilization of rainfall and reduction in the incidence of pests and diseases. A number of demonstrations have been taken up in farmers fields through ORPs, KVKs and IVLP programmes in different rainfed regions of the country. In case of sorghum and castor in farmers fields of Hyderabad, a fifteen day delay in sowing led to reduction of 300 and 850 kg/ha compared to normal sowing. Inadequate availability of farm implements and draft are major constraints. However, seeding and interculture experiments developed by CRIDA and AICRPDA centres helped in overcoming the constraints to some extent.

4. Adoption of improved crop varieties

A number of improved varieties and hybrids were evaluated in the farmers fields to identify suitable ones for matching growing periods for inter and sequence rainfed cropping systems. For example, farmers gained additional benefit ranging from Rs. 2000-4000/ha by adopting improved varieties of sorghum, castor and sunflower in Alfisols of Hyderabad.

5. Efficient crops and cropping systems

To achieve appropriate land use, efficient inter and sequence-cropping systems were recommended based on soil type, rainfall and length of growing seasons.

The studies at Hyderabad indicated only 25 percent farmers' adopted 2:1 ratio of sorghum-pigeonpea. Whereas 45 percent of farmers adopted the finger millet + pigeonpea system (8:1) ratio in Alfisols of Karnataka and maize + soybean system (2:2) was accepted by Ranchi farmers. Groundnut + pigeonpea (7:1) was widely accepted by the farmers in Rayalseema of Andhra Pradesh. Some of the constraints for wider adoption by the farming communities are 4 preference for fodder genotypes in cereals rather than grains for feed to live stock, lack of suitable farm implements to seed in different ratios, delay in planting of kharif for double cropping systems. These have to be refined under on-farm situations for greater acceptance by the farmers.

6. Farm implements

Proper tillage and precise placement of seed and fertilizers in the moist zone are most critical for successful crop establishment in drylands. Since the sowing of crops must be completed in a short span of time, use of appropriate implements is necessary to cover large area before the seed zone dries out. Suitable implements have been recommended for various locations to meet this requirement. These are designed to suit the soil type, crop and the draught power availability. In many cases, the existing local implement used by the farmers have been improved to increase their working efficiency. Studies at CRIDA in farmers' fields of Telangana indicated that use of the drill plough for sowing of castor and sorghum crops showed no variation in yields of the crops and plant as compared to farmers practice resulted 1½ times more coverage compared to farmers' method of seeding. Two labours who are required for placement of seed and fertilizer in farmers method can be saved with the drill plough. Thus a saving of Rs. 187/ha is possible with a drill plough compared to the traditional plough and plant system.

7. Nutrient management

Fertilizer recommendations in rainfed crop production have been made primarily for NPK along with the conjunctive use of chemical, organic and bio-fertilizer. Inclusion of legumes in cropping systems can supplement fertilizer N to the extent of about 20 kg N per ha. Conjunctive use of fertilizer N with FYM, croppings of luecaena and gliricidia help in reducing the requirement of fertilizer by 50 percent.

8. Integrated pest management (IPM)

Pests and deceases constitute a major constraint to increased food production. Crop losses due to pest attack range from 10-30 percent depending on the crop and environment. Complete crop failure may occur in case of serious attack. Indiscriminate use of the pesticides in rainfed crops will lead to harmful side effects such as direct toxically to the applicator or consumer, development of strains or pests resistant to pesticides, resurgence of pest species, outbreak of secondary pesticides, destruction of non-target organisms such as parasites and predators and accumulation of harmful residues of food products. Integrated pest management is one of the alternatives for the chemicals used for pest management. IPM encourages the most comfortable and ecologically sound combination of available pest suppression techniques and to keep the pest population below economic threshold. Easily adaptable and economically viable integrated pest management strategies have been developed for the control of major pest in rainfed crops like cotton and pulses.

9. Alternate landuse systems

Despite evolving a number of production technologies, arable cropping in drylands continues to suffer from instability due to aberrant weather. To provide stability to farm income and also utilize the marginal lands for production of fodder, fuel wood and fibre, a number of alternative land use systems were evolved based on location specific experimentation and cafeteria studies. In addition to the above general guidelines, specific experiments have been carried out to develop land use practices for different categories of soils across the centres integrating annual crops with the perennial component in order to utilize the off-season rainfall. Different alternate land use systems include agri-silviculture, silvi-pasture, agri-horticulture, alley cropping etc.

10. Integration of livestock with rainfed farming systems

Live stock is treated as a part of farming system in rainfed agriculture in India. The soil, plant, animal cycle is the basis for all feed used by the animals. The livestock in the rainfed regions are weak. Farmers in this area often sell their cattle due to the scarcity of fodder. In India the land holdings are being reduced with increased population pressure. Hence, land not suitable for agriculture has to be diverted for raising fodder need of animals through the appropriate alternate land use system such as improved pasture, silvipasture, hortipasture and tree techniques.

11. Integration of the technologies through watershed approaches

The concept of watershed is important in efficient management of water resources. As the entire process of agricultural development depends upon the status of water resources, the watershed with distinct hydrological boundary is considered ideal for taking up a development programme. In brief, planning and designing of all soil conservation structures are carried out considering the peak runoff. In this context, the watershed concept is of practical significance. Also, the entire development needs are to be taken up on topographic considerations from ridge to valley.

12. Resource conservation measures

Based on the nature and type of barriers and their cost, the conservation measures in arable lands can be divided into three categories: (i) Hardware treatments (ii) Medium software treatments and (iii) Software treatments.

13. Farming system approach

Of late, it has been increasingly recognized that unlike irrigated areas, it is difficult to develop profitable technologies for heterogeneous agro-ecological and socio-economic conditions of small holders in arid and semi-arid regions. Since, the problems are complex ,addressing only a component of the farming system, e.g crop variety, fertilizer use or even crop husbandry per se is not expected to bring about a significant increase in the productivity as witnessed in irrigated areas. The extension strategy should be such as to match this challenge. The farming systems perspective, developed on watershed approach therefore can be the appropriate management strategy for such regions.

The following steps constitute the farming systems mode for research, both on station and on-farm (Watershed)

- PRA and assessment of socio-economic conditions of people.
- Identification of ITK (indigenous technical knowledge).
- Collection of available technological knowledge on various components of the farming system – arable farming, animal husbandry, water harvesting, management of wastelands and alternate land use systems etc.
- Focus group (farmers) interaction to identify appropriate technology for different categories of farmers.
- Identification of lead farmers to function as facilitator in technology application and adoption.
- Identification of points of synergy among systems components.
- Structuring of technological components with maximum synergy.
- Phasing of program over the project period.

Chapter 8

Contigent Crop Planning for Aberrant Weather Conditions

Crop plan and practices for rainfall scarcity prone areas

The appropriate crop plan and practices, which have so far been generated and recommended for achieving the objective of increased and stable crop production in dry land areas have been summarized in the following points:-

1. Proper and timely tillage
2. Early sowing
3. Choice of suitable crop/good plant material
4. Proper plant population
5. Cropping system
6. Fertilizer use
7. Timely weed control
8. Efficient rain water storage and its use
9. Alternate land use system
10. Planning for aberrant weather hazard

1. Proper and timely tillage

Primary tillage is done in order to increase water intake and minimize weed infestation. Farmers do it by country plough. It has to be carried out as early as possible, prior to the sowing season. Keeping this in view, the concept of off season tillage has been developed. The practice increases soil moisture intake and results in efficient weed control.

Shallow off season tillage with pre monsoon showers ensures better moisture conservation and lesser weed intensity. In light soils, surface crust formation reduces seedling emergence and rainwater intake, thereby leading to more surface runoff A light cultivation would be adequate to open such soils. Area with hard pans which tends to develop hard pans due to consistent cultivation and Areas with textural profiles i.e. with more clay, as depth increase requires the deep tillage. Deep tillage helps in increasing water in soils having textural Profiles and hard pan. For in situ moisture conservation, land has to be opened so that it can cause hurdle to flow of rainwater. Tillage machines of appropriate size and type matching the power sources need to be used.

2. Early sowing

Dry land crops have to be planted at the earliest opportunity after the onset of monsoon. Under some situations, dry sowing is desirable. Early planting helps in better crop yield by providing better moisture conservation, good seedling vigour and availability of longer growing season and avoidance of pests and diseases.

3. Choice of suitable crops/good plant material

Selecting suitable crops and varieties capable of maturing within actual rain fall periods will not only help in enhancing production of a single crop but also in intensifying the cropping intensity. The capacity to produce a fairly good yield under moisture stress condition is the most desirable criteria for selecting a crop variety for dry lands. In other words, crop varieties for dry land areas should be of short duration, drought tolerant and high yielding which can be harvested within rainfall periods and have sufficient residual moisture in soil profile for postmonsoon cropping.The depth of soil, available profile moisture and amount and distribution of rainfall are some of the important factors playing an important role in the choice of crops.

4. Cropping systems

The rainfed areas are generally mono cropped having a cropping intensity of 100. Farmers either raise crop in Monsoon season or leave the field fallow in post monsoon season or conserve rainwater in monsoon and take a crop in post monsoon season on conserved moisture. However the researchers suggested that the cropping intensity can be increased through inter cropping or double cropping depending upon the weather situation taking the rainfall, soil type and other meteorological parameters into consideration, "effective growing season" have been worked out for different seasons. Based on this information, potential cropping systems have been suggested for the country.

Potential cropping systems in relation to rainfall and soil type

Rainfall (cm.)	Soil Type	Effective Growing season (weeks)	Cropping system
320-600	Red & Shallow Black soils	20	Single Kharif cropping
350-600	Deep Sierozems and alluviums	20	Single cropping in either kharif or
350-600	Deep Black soils	20	Single Rabi Cropping
600-750	Red, Black and Alluvial soil	20-30	Inter cropping
750-900	Alluviums, deep Black, Deep Red and Sub mountaneous soils	>30	Double cropping with monitoring
>900	Alluviums, deep Black, Deep Red and Sub mountaneous soils	<30	Double cropping assured

Double cropping

Double cropping is possible in the areas receiving more than 750 mm rainfall with a soil moisture storage capacity of more than 20 cm of available water for a successful double cropping. Early planning of the Kharif crop and its harvest as early as possible helps in quick establishment of the Rabi crop. Most of the Rabi crops cannot be planted because of Thermal sensitivity. The surface soil starts drying up after the harvest of kharif crop and seeding of subsequent Rabi crop sometimes poses Problem. In such situation, pre sowing irrigation for promoting seedling emergence might be important. Use of improved seeding devices like ridger seeder in dryland areas is beneficial.

Double	Cropping System for Hisar Region
Cluster bean	Bajra
Cluster bean	Gram

Inter cropping

In the seventies, agriculture became more production oriented after the availability of high yielding, short duration and inputs responsive crop cultivars. This developed the concept of inter cropping to produce more over space and time. Keeping this in view, research efforts have been made to identify the most remunerative inter cropping system for different regions in the country. While doing so due importance is given to most common crop of the region and to pulses and oil seeds.

5. Use of fertilizers

Soils of drylands in the country are not only thirsty but hungry also because these soils are severely eroded horizontally as well as vertically when ever efforts are made towards bunding and leveling of fields in dryland areas, it is the surface soil, which is removed. The result and effect is that the fields are rendered shallow in depth and completely deprived of plant nutrients, particularly N, P, K. It is therefore necessary to apply all the three major nutrients in adequate amounts. Fertilizer use plays an important role in drylands. Not only that its use insulates crops against moderate moisture stress as the root system would be more developed compared to unfertilized crops.

Thus fertilizer use helps in more efficient use of moisture, the scarce commodity in dry lands. Since soil moisture is limiting in drylands, the availability of nutrients becomes limited attempt should always be made to apply fertilizer drills drawn by bullock or tractor are available, the objective can be fulfilled. There has been belief among the farmers of dryland areas that use of fertilizer increases the chances of crop failure but recent findings have shown that the use of fertilizer is not only helpful in providing nutrients to crop but also helpful in efficient use of profile soil moisture. In dry land areas, a proper mixing of organic and inorganic material would be desirable. Organic forms have low nutrient content, but help to improve the moisture holding capacity of soils. In addition to yield advantage, nutrients like Potassium help to increase drought tolerance by affecting plant soil water relationship. Transpiration losses are reduced and productivity per unit water increases. For the maximum benefit from the applied fertilizer. There is proper placement and split applications of fertilizers need for maximum fertilizer use efficiency, which can achieve.

Fertilizer Recommendation For Dry Region Especially For Hisar Region

Crop	**Nutrients (kg/ha)**		**Method of Application**
	N	**P_2O_5**	
Bajra	40	20	In 2 splits, ½ drilled at sowing + ½ at knee high stage
Cluster bean	0	30	Placement
Green gram	20	40	All basal drilled
Chick Pea	15	40	Drilled
Mustard	40	20	Drilled

Note : 40 kg P_2O_5 ha to Chickpea if grown after Bajra. If sown after Kharif fallowing adjusts P_2O_5 based on soil test.

6. Timely weed control

Off season tillage, proper seed bed preparation and timely sowing help the crop in combating weeds. Studies conducted in the project very clearly show that the first 25 days period in case of 100 days crops and 35 days period in case of 150 days crops is sensitive to weeds. Therefore, one or two weeding with blade harrow or country plough keeps the weeds well in control. The effects are still better when the practices of inter culture operation with bullock drawn implements along with hand pulling of intra raw weeds is adopted and thus increase crop yields appreciably. For this purpose some implements like blade harrow, rotating weeders like dryland weeder are recommended. Though the herbicides are quite effective in controlling weeds but the high cost of these chemicals takes them beyond the reach of dryland farmers.

7. Rainwater management

Two approaches are being adopted in rainwater management viz. (a) on farm rainwater management and (b) harvesting rainwater into dug out farm ponds and recycling on the donor area.

a) On farm rainwater management

The entry of rainwater into the soil is through the surface. Therefore soil surface need be more receptive for rainwater. For this we do inter row water harvesting, inter plot water harvesting, sowing across the slopes and use of broad bed & furrows (BBF).

i) Inter row water harvesting : Cultivation of dry land of crops under ridge furrow system not only helps to take a good crop stand but also helps to conserve soil effectively. In this system, the raindrops absorbed where it falls and does not flow out of the field.

ii) Inter plot water harvesting : In dryland areas where rainfall (< 300 mm) is hardly adequate to meet water requirement of any of the dryland crops. Crops should be cultivated only in half of the land and the remaining half should be left uncultivated with provision for contributing runoff water to the cropped area. Crops cultivated on this basis are bound to yield fairly even in dry years. This technique is very good for cultivation for arid fruit trees like Ber etc.

b) Harvesting rainwater into dugout farm ponds

Dryland areas do receive atleast one heavy rainfall during monsoon season, which causes runoff particularly from the fields having undulating topography and crusting soil. Runoff in drylands varies from 1040 percent of rainfall. So effort has to be made to harvest atleast part of this run off in dugout ponds. To check the seepage losses from these ponds several sealing materials have been tried and cheap & effective materials have been identified e.g. in light soils, cement: soil moisture of 1: 8 or plasting with thick coaltar spray on the surface area of the pond are fairly effective.

c) Recycling of stored water

Even though moisture stress at any crop growth stage can reduce the yield, there are some stages, which are more critical than other. As the water availability is meager in dry lands so it is essential to apply it at right time to get the best return. So the critical stage for irrigation has to be chosen e.g. for rainy season sorghum the flowering stage is most critical and crucial. In Tobacco irrigation 3 weeks before topping had the maximum pay off. For wheat and barley crown root initiation stage is most critical stage. Long duration crops with deeper root systems that responded to higher levels of irrigation compared to short duration shallow rooted crops. This shows that it is important to wet the active root profile for best results.

8. Alternate landuse system

All drylands are not suitable for crop production. Some lands may be suitable for range / pasture management, while others for Tree farming, Ley farming, Agro forestry System. All these systems that are alternative to crop production are called as alternate land use system. This system not helps in generating much needed off season employment in monocropped drylands but also minimize risk, utilizes off season rains which may other wise go waste as run off, prevents degradation of soils and restores balance in ecosystem.

Crop production may be disastrous in the years of drought where as drought resistant grasses and trees could be remunerative. Scientists of dryland have developed many alternate land use systems which may suit different agro ecological situations. These are alley cropping, agri horticultural system and silvi pastoral systems that utilize the resources in better way for increased and stabilized production from drylands.

9. Planning for aberrant weather conditions

Aberrant weather is a common feature of rainfed agriculture. Four important aberration in the rainfall behaviour have been more commonly observed are:

i) The Commencement of rains may be quite early or considerably delayed.

ii) Dry Spell Immediately after Sowing.

iii) There may be prolonged breaks during the southwest monsoon season during which most of the dryland crops are grown and

iv) Rains may terminate earlier than normal cessation date or may continue beyond the normal rainy season.

i) Late onset of monsoon

During some of the years the onset of the southwest monsoon gets delayed so that the crops/varieties which are regularly grown in the region can not be sown in time. Delayed sowing of crop can lead to reduced and even uneconomical crop yields. Under these circumstances, two management options are available.

a) Transplantation : In case some of the crops, community nursery has to be raised at a point where water is available and transplantation has to be under taken with the receipt of rains. Under delayed conditions, transplanting of pearl millet has been found to be advantageous as the transplanted crop give significantly more grain yield than the direct seeded crop.

b) Alternate crops / varieties : Certain crops are more efficient either due to the shorter duration or capacity to produce better yields even under relatively unfavourable moisture regimes when sown late in the season. Therefore, crops/varieties have to be chosen depending upon the date of occurrence of sowing rains e.g. in western Rajasthan short location crops like green gram and cow pea were found to perform better than pearl millet under rate sown conditions.

Choice of crop varieties for normal & late sown condition

Crop	Centre	Normal Sown	Late Sown
Bajra	Hisar	HHB 50, HHB 60, HHB 67, HHB 94, HHB 117, HHB 197, HHB 216	HHB 67
Mung	Hisar	Muskan, S-9, SML 62,S 9HM 42,	Pantmung I
	Jodhpur	S 8	S 9
	Varansi	Varsha or K 851	T 44
Cow Pea	Hisar	FS 68, HFC 42-1	-
	Jodhpur	FS 68	K 11
Guar	Hisar	HG 365, HG 75,HG 563	FS 277
Sorghum	Akola	CSH 5, CSH 9	CSH 1
Mustard	Hisar	RH 30, Varuna,RH 781, RH 819,	RH 30
Sunflower	Hisar	EC 35737,27250	T 27
Chickpea	Hisar	GNG 146, C 235, HC 1	—

ii) Dry spell immediately after sowing

During some of the years a dry spell may occur immediately after sowing the crop. This may result in poor germination due to soil crusting, withering of seedlings and poor establishment of crop stand. It is always necessary to maintain proper plant stand to ensure better yields. Therefore, if severe dry spell occurs immediately after sowing it is better to resow the crop than to continue with inadequate plant stand.

iii) Break in monsoon mid or late season drought

a) Ratooning and thinning : The water requirement of crop depends upon the solar radiation intercepted by crop canopy. If the drought conditions occur

40-50 days after sowing, the leaf area development will be maximum leading to fast depletion of soil moisture reserve. Therefore reduction in leaf area index by either ratooning or thinning of the crop mitigates the ill effects of drought to a certain extent. Ratooning of a drought affected sorghum crop with a subsequent rain gave 8q/ha of grain yield over 2q/ha without ratooning. Ratooning is a high management technique and success depends on the general vigour of the drought affected crop.

b) Mulching : If the break in monsoon in very brief, soil mulching was found to be a tool in extending the period of storage of water in soil & hence the evaporation losses which in turn leads to extended periods of water availability.

c) Weeding : Weed control is very essential to save the crop from drought because weeds remove the soil moisture as well as nutrients.

d) Intercropping and risk distribution : The agricultural drought caused due to prolonged breaks in the monsoon rain can be of different magnitudes & severity and affect different crops in varying degrees. Application of meteorological information in ten n of frequency and probability of these breaks can be made to select a combination of crops of different duration in such way that there is time lag in the occurrence of their growth for appropriate intercropping.

   For the areas with uncertain rains in the early part of the season:early planting of deep rooted drought tolerant crops like Pigeonpea might be useful followed by planting of other component crop.

   For the areas with uncertain rains in the later part of season:the companion crop should be shorter in duration then the base crop e.g. Sorghum Green gram.

   In areas where rainfall is more or less uniformly distributed:It would be ideal to take suitable base crop and a companion crop of either longer or shorted duration depending on the growing season.

   For Hisar region, to minimize the risk the scientists suggested that 60 % area for Kharif Crops and Remaining 40 % should be allocated for Rabi crops and for successful Kharif System the 50 % area for Bajra, 25 % under Guar and remaining 25 % area should under short duration pulse/ forage crops.

e) Fertilizer use helps the Crop to Guard against drought by encouraging development of a root system, which will utilize soil water more efficiently.

iv) Early withdrawal of monsoon

This type of situation is more dangerous in drought prone areas. Early withdrawal creates difficulties of two types.

Sowing of Rabi crops may be jeopardized (are not sown in early Sept.)

Moisture conserved in the soil is required to be carefully used by appropriate practices.

The suggested practices are as under:

a) Reduced plant density : Rabi Jowar sown during early September at 1.0 to 1.35 Lakh plants per ha. require 50 % reduction. Plant population needs to be adjusted before plants go for their grand period of growth (30 35 DAS).

b) Use of surface mulch : Moisture can be conserved by using organic surface mulches. For this purpose, mulch material at five tones per ha. is required.

c) Protective irrigation : If possible protective irrigation may be given. Usually protective irrigation is propose to be given at 55 to 56 days growth. Due to early withdrawal of monsoon, the same may be applied at 35 to 40 days' growth.

d) Increase frequency of intercultivation : Early stoppage of rainfall brings early cracking in soil. To prevent cracking and there by loss of moisture, frequency of intercultivation may be increase. Usually 3 intercultivation are recommended, if same are increased to 5 or 6, it would be helpftil in creating dust/soil mulch.

e) Stripping of leaves : It helps to control moisture loss temporarily. This was not found to be useful practice for prolonged drought.

Improved dryland technology for rainfall scarcity areas

Following are the various improved techniques and practices recommended for achieving the objective of increased and stable crop production in dryland areas specially under rainfall scarcity areas.

i) **Crop Planning:** Crop varieties for dryland areas should be of short duration tolerant and high yielding which can be harvested within rainfall periods and have sufficient residual moisture in soil profile for post monsoon cropping.

ii) **Planning for weather:** Variation in yields and output of the dryland agriculture is due to the aberrations in weather conditions especially rainfall. An aberrant weather can be categorized in three types viz.

   a) Delayed onset of monsoon.

   b) Long gaps or breaks in rainfall and

   c) Early stoppage of rains towards the end of monsoon season.

   Farmers should make some changes in normal cropping schedule for getting some production in place of total crop failure.

iii) Crop Substitution: Traditional crops/varieties that are inefficient utilizer of soil moisture, less responsive to production input and potentially low producers should be substituted by more efficient ones.

iv) Cropping Systems: Increasing the cropping intensities by using the practice of intercropping and multiple cropping is the way of more efficient utilization of resources. The cropping intensity would depend on the length of growing season, which in turn depends on rainfall pattern and the soil moisture storage capacity of the soil.

iv) Fertilizer use: The availability of nutrients is limited in drylands due to the limiting soil moisture. Therefore, application of the fertilizers should be done in furrows below the seed. The use of fertilizers is not only helpful in providing nutrients to crop but also helpful in efficient use of soil moisture. A proper mixture of organic and inorganic fertilisers improves moisture holding capacity of soil and increase tolerance to moisture stress.

vi) Rain water management: Efficient rainwater management can increase agricultural production from dryland areas. Application of compost and farm yard manure and raising legumes add the organic matter to the soil and increase the waterholding capacity. The water, which is not retained by the soil,, flows out as surface runoff. This excess runoff water can be harvested in storing dugout ponds and recycled to donar areas in the severe stress during rainy season or for raising crops during winter.

vii) Watershed management: Watershed management is a approach to optimize the use of land, water and vegetation in a area and thus, to provide solution drought, moderate floods, prevent soil erosion, improve water availability and increase fuel, fodder and agricultural production on a sustained basis.

viii) Alternate Landuse: All drylands are not suitable for crop production. Some lands may be suitable for range/ pasture management, for tree farming dryland horticulture, agro forestry systems including alley cropping. All these systems that are alternative to crop production are called as alternate land use systems. This system helps to generate off season employment monocropped dryland and also, minimizes risk, utilizes off season rains, prevents degradation of soils and restores balance in the ecosystem. The different alternate land use systems are alley cropping, agri horticultural systems and silvi pastoral systems, which utilize the resources in better way for increased and stabilized production from drylands.

Efficient management of rainfed crops

Dryland farming refers to high yield and high efficient agricultural production in areas without irrigation capability and mainly depending on natural rainfall through the adoption of a set of dryland farming practices.

Major measures

The objectives for the development of dryland agriculture are as Follows: (I) to enhance moisture retention capacity of the soil and retention rainfall by using agronomic, biological and engineering measures in an integrated way; (II) to improve soil fertility and conserve land (soil, water and fertilizer) (III) to change farming practices and make full use of resources like light, heat, soil, fertilizer, water and improved seeds to increase yield and agricultural productivity in the dryland. Major measures include:

Terracing

This measure is meant to build slope land with an angel less than 25 degrees into contour terrace. Slope land with an angel of 10 25 degrees is susceptible to soil erosion due to large angel, steep slope, frequent farming activities and high cultivation coefficient, particularly improper farming practices. Investigation results have shown that farming activities on such slope land have each year led to a soil erosion of 0.43 cm of soil layer and loss of 48 tons of surface fertile soil per ha. To change such slope land into contour terrace, plus other measures like small catchments improvement and biological and agronomic practices, will help to improve production conditions, prevent water and soil erosion and raise land quality and grain yield. Slope land will be improved to plant crops along the contour using trench culture with 1/2 meter deep and I meter wide trench or ridge culture featuring three-dimensional farming approaches with improved soil so as to conserve soil and water, improve soil fertility and facilitate sustainable development.

Furrow drilling

This measure means to use farming practices like moisture retention mulched furrow, machine furrow for drilling and large furrow. The main purpose is to increase active soil layer, improve moisture retention capacity and soil fertility, reduce soil evaporation and improve ecosystem. The Yanan practice has shown that such measure can increase soil moisture retention by 81.87 cubic meters per mu in places with annual rainfall of 197 mm. This technique is suitable for any slope land with annual rainfall over 400 mm, or for terrace land. When it is used for small angel slope land and terrace land,, machines or animals can be employed to reduce labour intensity and speed up engineering progress. Presently, this technology has been used in 10 million ha in northern Shaanxi, southern Shaanxi, eastern Gansu, northern Hebei and northwestern Shanxi, resulting in yield increase of around 40 60%.

Water cisterns

This measure is meant to build water cisterns to collect rainfall as supplementary irrigation water. In slope land and terrace land, water retention works will be built to collect natural rainfall as supplementary irrigation water for agriculture. In case of serious drought, such water can be used for drip irrigation to increase soil moisture. Due to limited water volume, such technology is usually used together with other water saving measures like wet sowing, plastic mulching, root zone drip irrigation, hole irrigation with mulching so as to enhance crop resistance to drought and ensure stable, high yield. It is applicable to places with annual rainfall less than 350 mm. To build one water cistern with a capacity of 50 cubic meters need an investment of around 400 yuan. It can provide supplementary irrigation water for 2 ha of land and ensure a yield per ha of over 300 kg. If used in production of cash crops like water melon, greenhouse vegetable and fruit tree, it will bring about even greater yield increase.

Mulching

This refers to mulching with plastic film and crop residue. It can reduce moisture evaporation, increase moisture retention capacity of the soil, alleviate the threat of drought and improve water use efficiency. Plastic film can be used for hole sown wheat, rice, filmside sowing, multiple crops using the same mulching, mulched corn field for water collection with very good results. Wheat farming with plastic film represents a major breakthrough in plastic mulching technology, contributing tremendously to the reform of dryland farming practices and increase of wheat yield. Experiments have shown that wheat with plastic film can produce 30% more than conventional wheat drill seeding,, increasing yield per ha by 100 kg, net income by over 100 yuan and saving water by 100 cubic meters. Presently, it is being tried and extended in provinces like Hebei, Gansu, Shaanxi, Ningxia, Qinghai, Inner Mongolia and Shandong. Mulching with crop residue features easily accessible material low cost of high efficiency,, water saving,, moisture retention,, fertility enhancement,, yield increase and no contamination to soil. It can usually improve soil moisture retention rate by 30 50% and increase water supply by 40 80 cubic meters per mu. Plus balanced fertilization technology like nitrogen, phosphate and potash replenishment, it can at least improve yield per mu by 100 kg, representing an effective solution to the problem of dryness and 5 infertile soil in dryland region.

Coating

It refers to seed coating mainly using drought resistant agent. The occurrence of drought is often unpredictable. Conventional measures can hardly play their role under such circumstances. Drought resistant chemical agents developed in recent years, like water retention agent, evaporation suppressant and soil regulator, can bring very good results at low cost when used at a time of drought. For instance, seed coated with water retention agent can improve moisture use efficiency by 0.05-0.17 kg/mm; compared with the control group, wheat seed coated with drought resistant agent can increase yield by 20%; and if sprayed before the xerothermic wind, it can increase yield by 9.5-18%.

Fertilization

It refers to the application of chemical fertilizer, organic fertilizer and green fertilizer to improve soil fertility and yield. Besides drought, another factor constraining productivity in dryland farming area is soil infertility. By moisture control through fertilization, we mean that we can improve soil resistance to drought through the improvement of soil fertility. The main ways include:

1. To establish rational and effective dryland farming structure, increase the percentage of legume and green fertilizer crops and combine farming with fertility improvement. The key is to adopt a grain crop and summer green fertilizer rotation system or grain oilseed legume rotation system as well as other farming models like grain legume and grain grass intercropping and wheat potato intercropping in order to fully use natural resources like light, heat and water to achieve yield and income increase.

2. To increase fertilizer input by combing organic fertilizer with chemical fertilizer and improve scientific fertilization level. Practices have shown that increased use of chemical fertilizer on dryland can bring about twice as much yield as on irrigated land. Meanwhile, the use of technology like crop residue mulching and quick rotting of crop residue as fertilizer can also increase the application of organic fertilizer and improve the use efficiency of chemical fertilizer.

Deep tillage

This refers to mechanized deep ploughing technology. Through deep ploughing by machines, we can break the subarable layer without disturbing surface soil layer so as to improve soil ventilation and rainfall retention capacity. This cannot only be used for dryland but also for irrigated land.

Seed replacement

This refers to the use of drought enduring and drought resistant seeds. To use drought enduring varieties is the most cost effective yield increasing technology, At present, most dry areas have their own varieties with good drought resistance. However after such a long time,, most varieties have experienced degradation in their performance. And the breeding of new varieties is still lagging behind. It is an urgent need to develop drought resistant varieties and accelerate the purification and rejuvenation process in order to improve yield level.

Drought management strategies

In our country, irrigated area constitutes 33% and 67% is dryland and rainfed, out of 143.8 mha of cultivated land. Total cropped area is 180.36 m ha. Such dryland agriculture contributes about 45% of national food grain production. After full exploitation of dryland, it may contribute up to 75% of total food grain production. Pulses and oilseeds are mainly grown in such areas. Important commercial crops viz. cotton, castor groundnut and all coarse grains viz. jowar, bajra, maize crops are rainfed. Problems of dryland agriculture are inadequate, uncertain, uneven rainfall. Drought is major problem, so we go for contingency planning.

Nature of drought

Haryana receives almost 1200-1300 mm of rainfall every year but the rains are erratic in many areas. The probability of rainfall failures and coefficient of variations is quite high in the last weeks of June-July and in the last weeks of September-October. Hence, drought in the state primarily occurs at the start or end of the Kharif season. In July, upland crops grow to maturity and seedlings for transplanted rice are established. If there is deficient rain, the upland crop, mainly paddy and maize, which provides food security in August-September, is affected. Seedlings for the transplanted paddy start to wilt or become over-mature.

As most of the land is monocropped, the lowland paddy is crucial for employment and food security. A delay in rainfall affects the transplantation of paddy, the major crop in the area. October rains are required for paddy and provide the necessary residual moisture for the Rabi crop. The entire state is not generally affected by drought. However, there are areas which are affected frequently. Over a period of time, new areas become affected by drought, and we see that drought has been officially declared in areas outside the DPAP.

When the rains fail, agriculture is usually the first to be affected because of its critical dependence on stored soil water. First, soil water in the uplands starts to deplete. Then shortage of water starts to affect people collectively and individually.

The socio-economic impact of drought occurs sooner as frequent droughts have weakened the capacity of the people to bear shock. While drought is basically caused by erratic and deficient rain, the problem has been aggravated by large-scale open-cast mining; deforestation; irregular and non-scientific mining and quarrying; inefficient management of resources, particularly water resources; and decline in traditional systems of water management.

The different strategies for drought management are discussed under the following heads.

- Adjusting the plant population: The plant population should be lesser in dryland conditions than under irrigated conditions. The rectangular type of planting pattern should always be followed under dryland conditions. Under dryland conditions whenever moisture stress occurs due to prolonged dry spells, under limited moisture supply the adjustment of plant population can be done by
- Increasing the inter row distance: By adjusting more number of plants within the row and increasing the distance between the rows reduces competition during any part of the growing period of the crop. Hence it is more suitable for limited moisture supply conditions.
- Increasing the intra row distance: Here the distance between plants is increased by which plants grow luxuriantly from the beginning. There will be competition for moisture during the reproductive period of the crop. Hence it is less advantageous as compared to above under limited moisture supply.

Mid contingent plan for aberrant weather conditions

The contingent management practices done in the standing crop to overcome the unfavorable soil moisture conditions due to prolonged dry spells are known as mid season conditions.

- Thinning: This can be done by removing every alternate row or every third row which will save the crop from failure by reducing the competition
- Spraying: In crops like groundnut, castor, red gram, etc., during prolonged dry spells the crop can saved by spraying water at weekly intervals or 2 per cent urea at week to 10 days interval.
- Ratooning: In crops like sorghum and bajra, ratooning can practiced as mid season correction measure after break of dry spell.

Mulching: It is a practice of spreading any covering material on soil surface to reduce evaporation losses. The mulches will prolong the moisture availability in the soil and save the crop during drought conditions.

Weed control: Weeds compete with crop for different growth resources ore seriously under dryland conditions. The water requirement of most of the weeds is more than the crop plants. Hence they compete more for soil moisture. Therefore the weed control especially during early stages of crop growth reduces the impact of dry spell by soil moisture conservation.

Water harvesting and life saving irrigation: The collection of runoff water during peak periods of rainfall and storing in different structures is known as water harvesting. The stored water can be used for giving the life saving irrigation during prolonged dry spells.

Short term strategy of contingency planning

Immediate planning

- Judicious use of surface and groundwater for drinking and irrigation and districts.
- Ensuring availability of quality fodder to animals.
- Livestock management including establishment of fodder/feed depots and cattle camps.
- Selection of crops, cropping sequences and agronomic practices for drought affected areas.
- Promotion of subsidiary income and employment generating activities.
- Gainful implementation of NREGA, RKVY, NFSM, NHM, RGGVY, BRGF and other schemes.
- Deployment of Information Technologies.

Contingent cropping

- Date bound contingency crop plan for different meteorological sub-divisions and agro-ecological regions.
- Early maturing varieties of different Kharif crops.
- Crops and cropping system-wise agronomic practices.
- Technology for Rabi/summer Rice in Boro cultivation area.
- Potential districts for Rabi Maize.
- Soil & Water Conservation Measures for different rainfall regions.

Medium and long term strategy

- Securing Good Quality Water in Drought Prone Areas.
- Perennial and Non-conventional Fodder.
- Improved Live-stocking, Breeding and Management.
- Up gradation and Fine Tuning of Crops, Cropping and Farming Systems.
- Exploiting under-exploited and under-utilized plant resources.
- Creation of Alternate Income and Employment Generating opportunities.
- Time bound Rabi Plan based on receipt of rainfall in August and September.
- Resource conservation technologies including zero tilled specially in Indo-gangetic plains.

- Need based location specific agronomic manipulations.
- Resorting to cultivation of less water demanding crops and cropping systems in limited irrigation water supply areas.

Preparation of appropriate crop plans for dryland areas

Although, the past meteorological records help sufficiently in planning for future, but because, the drought is a phenomenon which may occur at any time during the crop period, so, we have to make our planning accordingly. All India Coordinated Research Project for Dryland Agriculture has evolved Several Contingent Crop Production Strategy to meet the challenges of acute drought conditions, which are discussed here under :

Modern practices

- Choice of suitable crops
- Choice of suitable crop varieties
- Cropping system (Intercropping) with
- Planting pattern
- Planting time
- Fertilizer management with dose, time and method of application.
- Seed hardening

a) Mid season corrections

- Alternate crop strategy
- Transplanting of pearl millet
- "Life-saving" irrigation
- Adjustment of plant population
- Leaf stripping

b) Moisture conservation practices

- Use of mulches
- Use of antitranspirants and light reflectants
- Interculture operation
- Weeding
- Water harvesting techniques
- Traditional Practices

- Contour building
- Ridge and furrow system
- Deep ploughing

At the household level, the strategies adopted before drought include:

- Crop diversification/ crop variety diversification
- Livelihood diversification
- Staggering of seedling periods
- Early migration
- Keeping land fallow
- Limited re-sowing.

Strategies adopted after drought sets in include:

- Increased re-sowing if the rains come
- Shift to pulses on upland and medium-level land
- Early sowing of rabi crop
- Providing irrigation to paddy seedlings
- Transplantation of over-mature paddy seedlings to get straw for feeding animals
- Migration
- Asset depletion
- Borrowings
- Reduction in consumption

In fragile zones and among the poorest households, reduction in consumption is ultimate coping mechanism.

Adaptations of crops to drought

For agricultural crops, climate-related natural disasters are the leading source of risk and uncertainty. One such climate-mediated event is drought, a situation where the amount of rainfall is substantially below average for a particular area. This can have a particularly negative effect on agricultural crops, which require ample water to grow successfully. Because plants must lose water when they take up carbon dioxide from the air, crops require 500 to 1000 grams of water for each gram of harvested product, Lack of moisture places the plant under stress and can significantly reduce crop yields. It is estimated that 38% of the world area, or 70% of the agricultural output, is affected by drought. Irrigation can markedly increase crop productivity, but lack of availability or delivery infrastructure and competing demands for water (urban, industrial, environmental) are pressuring fresh water supplies, making water deficits one of the greatest challenges for agricultural production Since drought patterns are often unpredictable, farmers have difficulty predicting yields and famines can occur in years when water supplies are inadequate. In many countries, drought is a major contributor to food insecurity and poverty.

Appropriate techniques for successful crop production in dryland areas

In every region of the world it is necessary to find or develop appropriate techniques for agriculture. A large part of the surface of the world is arid, characterized as too dry for conventional rain fed agriculture. Yet, millions of people live in such regions, and if current trends in population increase continue, there will soon be millions more. These people must eat, and the wisest course for them is to produce their own food. Yet, the techniques are so varied that only a very large volume would cover the entire subject. This publication is only a primer, an introduction to appropriate techniques. More extensive treatments are mentioned in the bibliography. In many cases the most suitable techniques for a particular region may be those already developed by the local inhabitants. In some cases it will be difficult to improve on local techniques, but at times even simple and inexpensive innovations may be almost revolutionary. This technical note suggests that one must begin to improve local agriculture in arid zones by learning what is already there. Then both techniques and plants that may be useful in specific situations are suggested.

Chapter 9

Watershed Management Definition, Objectives, Concepts, Approach, Components, Steps in Implementation of Watershed

Watershed

A watershed is the area of land where all of the water that is under it or drains off of it goes into the same place. John Wesley Powell, scientist geographer, put it best when he said that a watershed is: "that area of land, a bounded hydrologic system, within which all living things are inextricably linked by their common water course and where, as humans settled, simple logic demanded that they become part of a community."

Watershed is defined as a hydro-geological unit of area from which the rainwater drains through a single outlet. Watershed development refers to the conservation, regeneration and judicious use of all the natural resources (like land, water, plants, animals) by human beings. Watershed Management brings about the best possible balance between natural resources on the one side and human beings on the other. Human beings and the ecology are interdependent. The changes in the environment directly affect the lives of the people depending on it. A degraded environment means a degraded quality of life of the people. This degradation can be tackled effectively through the holistic development of the watershed. A watershed provides a natural geo-hydrological unit for planning any developmental initiative. The approach would be treatment from "ridge to valley".

A watershed, also called a drainage basin or catchment area, is defined as an area in which all water flowing into it goes to a common outlet. People and livestock are the integral part of watershed and their activities affect the productive status of watersheds and vice versa. From the hydrological point of view, the different phases of hydrological cycle in a watershed are dependent on the various natural features and human activities. Watershed is not simply the hydrological unit but also socio-political-ecological entity which plays crucial role in determining food, social, and economical security and provides life support services to rural people.

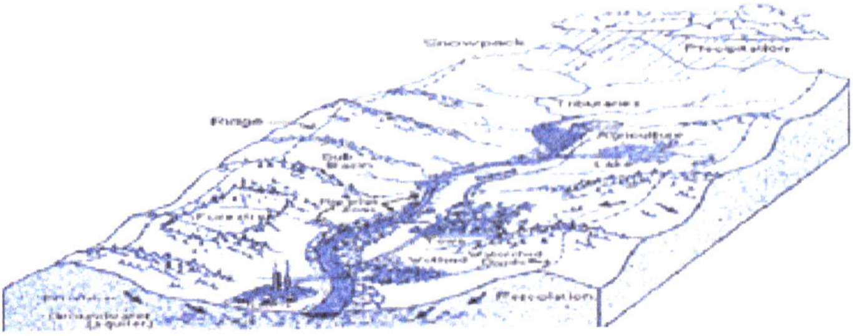

Entry Point Activity (EPA) *Source*: Produced by Lone Council of Government

Entry Point Activity is the first formal project intervention which is undertaken after the transect walk, selection and finalization of the watershed. It is highly recommended to use knowledge-based entry point activity to build the rapport with the community. EPA must be avoided as such activities give a wrong signal to the community at the beginning for various interventions. Details of the knowledge-based EPA to build rapport with the community ensuring tangible economic benefits to the community members are described here.

Objectives of watershed management

As mentioned earlier, watershed management is enshrined with the concept of sustainability meeting the needs of present population without compromising the interests of future generations. It is multi-pronged approach for steady uplift of masses living in the area. The main objectives of this multipurpose programme can be described in symbolic form by the expression: "Power". Here the letters symbolize the following:

P = - Production of food –fodder - fuel – fruit – fibres – fish – milk combine on a sustained basis
- Pollution control
- Prevention of floods

O = - Over exploitation of resources to be minimized by controlling excessive biotic interferences like over-grazing.
- Operational practicability of all on farm operations and follows up programmes including easy approachability to different locations in watershed

W = - Water storage at convenient locations for different purposes,
- Wild animal and indigenous plant life conservation at selected places.

E = - Erosion control, Ecosystem safety,
- Economic stability,
- Employment generation

R = - Recharge of groundwater
- Reduction of drought hazards
- Reduction of siltation in multi – purpose reservoirs
- Recreation

The above objectives can be achieved by planning and implementing the programme in a systematic way with active participation of farmers including constitution of Cooperative Watershed Management Societies rather, as a convention not as a rule following guidelines may be followed:

- The implementation programme should start from the ridge line of the watershed *to* the valley not on piecemeal basis in isolated patches.
- Development of both arable and nonarable lands should be done together.
- Forest, pasture, cultivable land and waste- lands should be treated as inter linked units of hydrological entity. The condition of all lands has to be improved *to* meet the demands of increasing man and animal population.
- Essentially, all developmental activities are to be carried out on watershed basis – whole watershed area needs to be covered – may be in planned phases. However, treatment of some lands lying outside the watershed but geographically contiguous can also be taken *up* based on socio economic considerations.

Objectives of watershed management can also be broadly expressed as:

1. To control damaging runoff and degradation and thereby conservation of soil and water
2. To manage and utilize the runoff water for useful purpose
3. To protect, conserve and improve the land of watershed for more efficient and sustained production
4. To protect and enhance the water resource originating in the watershed
5. To check soil erosion and to reduce the effect of sediment yield on the watershed
6. To rehabilitate the deteriorating lands
7. To moderate the floods peaks at downstream areas
8. To increase infiltration of rainwater
9. To improve and increase the production of timbers, fodder and wild life resource
10. To enhance the ground water recharge, wherever applicable

Principles of watershed management

The main principles of watershed management based on resource conservation, resource generation and resource utilization are:

1. Utilizing the land according to its capability;
2. Protecting productive top soil;
3. Reducing siltation hazards in storage tanks and reservoirs and lower fertile lands,
4. Maintaining adequate vegetation cover on soil surface throughout the year,
5. *In-situ* conservation of rainwater,
6. Safe diversion of excess water to storage points through vegetative waterways,
7. Stabilization of gullies by providing checks at specified intervals and thereby increasing ground water recharge;
8. Increasing cropping intensity and land equivalent ratio through intercropping and sequence cropping
9. Safe utilization of marginal lands through alternate land use systems with agriculture horticulture forestry pasture systems with varied options and combinations;
10. Water harvesting for supplemental and off- season irrigation;
11. Maximizing agricultural productivity per rage at convenient locations for unit area per unit time and per unit of water.
12. Ensuring sustainability of the eco system befitting the man-animal- plant water complex.
13. Maximizing the combined income from the inter related and dynamic crop livestock tree labour complex over years;
14. Stabilizing total income and to cut down risks during aberrant weather situations;
15. Improving infrastructural facilities with regard to storage, transportation and marketing of the agricultural produce;
16. Setting up of small scale agro industries; and
17. Improving the socio-economic status of the farmers.

The main components of watershed management are:

- Soil water conservation and water harvesting.
- Crop management
- Alternate landuse systems

Successful implementation of the above three components in the watersheds would lead to the development of livestock, poultry, pisciculture and other associated and allied activities, finally resulting in improved standard of living of the farmers and rural masses.

Soil water conservation and water harvesting

Soil and water conservation measures coupled with water harvesting help to improve the moisture availability in the soil profile and surface water availability for supplemental and off season irrigation. The interventions through soil conservation measures have greater role to play in transferring a part of surface water to ground water by recharge. Based on the nature and type of hydraulic barriers and their cost, the conservation measures in arable lands can be divided into following three categories.

1. Hardware treatments
2. Medium software treatments
3. Software treatments

Hardware treatments

Hardware treatments are generally of permanent type, provided for improvement of relief, physiography and drainage features of watershed. These are erected with major Government support with the purpose to check soil erosion, regulate overland flow and reduce peak flow. At times, they are imposed to completely divert the upland run off from running into downstream fertile lands. Different hardware treatments are explained below.

Waterways: Quite often, high run off volumes are observed, even in drylands, particularly in watersheds located in sloppy areas. Such run-off water should, therefore, be chanelised through a few waterways; as far as possible some of the existing waterways may be developed. Waterways draining larger areas may be designed on hydrologic and hydraulic considerations. No doubt, at places, sizeable land strip may be occupied by such waterways, but his may be found highly advantageous for the survival of may multi-purpose useful trees when planted along the waterways. Depending upon land situations alignment of small size waterway may be adjusted with field boundaries; at times, such arrangements are specially useful for safe guarding crops against over stagnation of rain water in black soil areas. Such water course may also be safeguarded against erosion by providing mechanical checks and also by raising vegetation which may be useful fodder for animals. In some salutations, waterways may also have a small section side bund with necessary openings at water entry points.

Bunds : These are low height earthen embankments constructed across the slope in cultivated lands after deciding location of waterways. The bunds function to intercept runoff, increase infiltration opportunity time and dispose excess rainfall safely. Such bunds can be either contour bonds or graded bunds; many a time, these bunds are adjusted with field boundaries, if deviation from grade or contour is not too much; no doubt, spacing depends on many engineering considerations.

Contour bunds are recommended in dry farming areas with light textured soils of slopes upto 6 per cent and where annual rainfall does not exceed 600 mm. They are designed for an expected run-off of 24 hours duration and 10 years frequency. Surplussing arrangements (waste weirs) are provided to dispose of the excess run off beyond the design storage. The cross section area of contour bunds follows the depth and type of soil; however, 0.5m^2 is minimum.

Graded bunds are constructed in medium to high rainfall areas in poor permeable soils (very-tisol), having 2 to 6 per cent slope. They are also quite suitable for the soils having crust formation tendency like red *'chalka'* soils of Telengana region of Andhra Pradesh. By and large, graded bunds with 0.3 to 0.5 in section are constructed with longitudinal gradient of 0.2 to 0.4 per cent depending on the site condition. Graded bunds particularly are made with the following objectives:

- To reduce run-off;
- To reduce soil loss; and
- To divert run off.

Thus, graded bunds along with waterways and water harvesting structures not only check soil erosion, but they also provide an ideal rain water management system for many watershed situations.

For making bunds, required soil may generally be taken from lower side. Sometimes, excavations and pits for making bunds in deep black soil areas particularly are made on upper side. But these may be got levelled or ploughed up as early as possible after bund constriction. Invariably barring a few situations borrow pits should be made on lower side at the places where the bunds drain into waterways. In real field situations, bunds are covered with natural grass and in most cases, these are dependable sources of forage. In addition, quite often some trees also comp up in their close vicinity, to the liking of cultivators. In some places loose stones are easily available; hence stone bunds are quite advisable in such pockets.

Terracing: Cultivated lands having land Slopes a 10 per cent particularly in hilly areas should be put under bench terracing by converting the lands into series s of platforms. The width of bench terrace depends on the land slope and the permissible depth of cut. Bench terracing is very much effective in reducing soil erosion in hilly areas. At Ootacamund, it has helped in bringing down the annual erosion rate from 39 t/ha to less than 1.0 t/ha on 25 per cent sloping lands. At places where scattered stones are available, loose stone walls can be made to act as risers for bench terrace construction.

Zingg terrace consists of a contributing (donor) area with mild slope and a receiving levelled area at lower side half to one third of the donor area. They are particularly suited for areas having medium to deep soils.

Medium software treatments

Medium softwares are also provided particularity as inter bund treatments, where field sizes are large and conventional bunds are con-structed along field boundaries. Such treatments are usually of semi-permanent type and are adopted to minimise the velocity of overland flow. They may need major initiative from cultivators in addition to some grants from the Government side. Such measures may last 2 to 10 years; vegetation component and land configurations at times may also provide some direct returns on short term basis. However, they need to be modified at times to maintain effectiveness for erosion control and moisture conservation, and also to minimise risk of providing shelter to harmful pests in and around these measures and in certain situations like taking up corrective measures to avoid too much spread of into-diversed vegetation and being counter productive.

Small section/key line bunds: A small section bund may be created across the slope at half of the vertical bund spacing. Such bunds can be nearly 0.1 m2 in section; may be renovated at an interval of 2 to 3 years.

Strip Levelling: About 4 to 5 in strip of land above the bund across the major land slope may be levelled for the purpose. Similarly, one or more strips can be created at midlength of slope. Such strips can be created by running blade harrows after ploughing the field with mould board or disc plough. Such minor rough levelling programme, may be taken up after every 2 to 4 years.

Live beds: One or two live beds (2 3 m wide) can be created either on contour or on grade in the inter bund space. The vegetation on the beds may be according to the liking of cultivator, it can be annual or perennial or a combination of both.

Vegetative/live barriers: One or two barriers of close growing grasses or legumes can be created along bunds as well as at the mid length of slop~ to filter the run off water or slow down, overland flow. *Khus* could be one of such vegetations. Several other promising grass species that serve as valuable fodder for cattle are being explored as an alternative to *khus* grass. A miniature bund at lower side of barrier is recommended to help in the development of live barriers, particularly in the initial stages.

Software treatments

A mention was made that hardware type land treatments are useful for safe run off disposal and similarly, medium soft wares are essential to slow down the velocity of overland flow in cultivated fields. However, on several occasions, these are found inadequate in attaining equitable moisture distribution for crop growth. In such cases software treatments are taken up for ensuring fairly uniform soil moisture for satisfactory crop performances. By and large, software treatments are temporary in nature; in that case these are required to be remade or renovated every year. The entire cost of applying such treatments is to be met by the farmers. Because of favourable economics, a few of these treatments have gained wide acceptance in the recent years.

Contour farming: Contour farming is one of the easiest and most effective and low cost method of controlling erosion and conserving moisture. With contour fanning, tillage operations are carried out along contours. It creates numerous ridges and furrows for harvesting sizeable amount of runoff inside the field itself.

Compartmental bunds*:* Compartmental bunds, converting the area into square/ rectan-gular parcels are useful for temporary im-pounding of water for improving Moisture status of the soil. These are made using bund formers. In medium deep black soil, they are found ad-vantageous in storing the rainfall received during the rainy season in the soil profile there- by augmenting the soil moisture. for use by rabi crops. The size of the compartments may be fixed' considering the size, of the inter bunded land.

Broad bed and furrows: Broad bed and furrow system furrows. This technique is especially suited to black soils, where crops are sown on pre formed beds. This system is made before the season and is maintained year after year. The planting is done on the bed. Generally, the depth of each furrow is kept 0.15 rn and the inter furrow spacing is maintained at 1.5 m.

Dead furrows: Dead furrows are laid across the land slope in rolling lands, to intercept the run off. The spacing between dead furrows varies between 2 to 5 m or 4 to 7 crop rows. This system works well in alfisols.

Tillage: Tillage operations help in rain water infiltration. Off season tillage, in particular, has been found quite useful in most rainfed areas. The tillage operations make the soil receptive to rainfall. This practice is very useful in light soils often prone to crusting.

Mulching: Surface mulching protects soil against beating action to rain drops and it also increases water infiltration into the soil. Further it helps in minimising water evaporation from soil surface. Sometimes dry soil mulch created simply by stirring the soil has been found effective for good performance of crops.

Soil and water conservation in non-arable lands

Control grazing is one of the simplest approaches in reducing soil loss from denuded sloping lands. As seen from the experiences of Operational Research Projects, soil and moisture conservation measures are required even in the non agricultural lands. These practices include contour and staggered trenches and contour furrows. They help in accelerating vegetation establishment, encouraging natural regeneration of species and combating soil erosion. Contour and staggered trenches are usually spaced 10 to 20 m apart across the slope for raising forest species. In general, 0.45 m wide trenches with 0.45 m depth are made at regular intervals. Sometimes in place of trenches, contour furrows are formed. At a spacing of 2 to 10 m. In rocky areas, crescent bunds of loose boulders are constructed in place of trenches. All the above stated measures are useful in slowly regaining the. lost forest cover and fertility of most, degraded grasslands; community efforts in the right direction are likely to make major dent in re greening such areas.

Water harvesting measures

Since time immemorial, several kinds of water harvesting structures have been in practice in our country. Recently, local experiences and precedence's were utilised for designing these structures. In order to bring cost effectiveness, efforts are also necessary to harmonize the designing of water harvesting structures based upon traditional wisdom and scientific contemplations. In3act, farm ponds, surface water tanks and percolation tanks need to be viewed as 'conservation structures in addition to sources for supplemental irrigation. In real watershed situations, ponds maybe located in conjunction with waterways some maybe of a few hundred cubic metre size, .whereas couple of structures may be of bigger size. In selected pockets, integrated catchments cum command approach should be followed. To ensure proper storage in light soils, some of the ponds can be lined with sealant materials. All large ponds have to be constructed at Government cost. All such structures are likely to recharge ground water reservoir. When located in proper co cultural hydrologic matrix, in conjunction with other factors, they also help in reducing load on ground water aquifer. It maybe advisable to work out design details of all the required structures in the first phase of the project, no matter construction of a few of these may have to be deferred may be planned to be taken up in drought years to provide employment to the rural folk. The objectives of water harvesting structures are:

- To store water for supplemental and off season irrigation;
- To act as silt detention structure;
- To recharge the ground water;

- To raise aquaculture species; and
- Recreation and allied agricultural uses.

Thus in most situations, in addition to reviving traditional water harvesting systems, other measures are also needed. Some of such typical water harvesting structures adopted in watersheds is as follows:

Minor irrigation tanks: *Minor* irrigation tanks are constructed across the major streams with canal system for irrigation purpose by con-structing low earthen dams. A narrow gorge should be preferred for making the dam in order to keep the ratio of earthwork to storage as min-imum. The height of the dam may vary from 5 to 15 m. The tanks are provided with well-designed regular and emergency spill ways for safety against side cutting. In micro watersheds, water harvesting *bundries,* similar to small scale irrigation tanks are also recommended. By and large, the water harvesting *bundhies* are not integrated with extensive canal system.

Farm ponds: Farm ponds are typical water harvesting structures constructed by raising an embankment across the flow direction or by excavating a pit or a combination of both. Dug out ponds in light soil require lining of the sides as well as the bottom with suitable sealants. This is not the case with heavy black soil. Considering the benefit of the harvested water, different type of linings, viz., brick, concrete, HDPE, etc. may be used.

Nala bunds and percolation tanks: Nala bunds and percolation tanks are located in the nalas having permeable formations with the primary objective of recharging the ground water. A strict regulation on the silt load entering the downstream reservoirs is an additional ad vantage of percolation tanks. The percolation tanks encourage the digging up of wells downstream of recharged area for irrigation purposes. The percolation tanks are provided with emergency spillways for safe disposal of flow during floods.

Stop dams: *Stop* dams are permanent engineering structures constructed for raising the water level in the *nala* for the purpose of providing life saving irrigation during drought periods. These are located over flat *nalas* at narrow gorges carrying high discharge of long durations. They are created over stable foundation conditions where hard murrum or rocks are encountered. For the stop dam, a site with larger water storage capacity would be desirable.

Crop Management

The rainfed areas of our country contribute about 45 per cent of the total food grain production. In the watersheds there is scope of increasing the overall productivity by adopting the integrated farming systems approach. For achieving sustainability in crop production in the watersheds due importance is given to : (i) evolve and follow up simple and low cost crop production technology including improved varieties, (ii) provide alternate crop production technologies to match weather aberrations; and (iii) optimize the use of natural resources like land and water.

Alternate landuse systems

In the watersheds, most of the uplands are degraded and these are also marginal lands with low productivity level. Due to the population pressure, more and more marginal and sub marginal lands are being brought under cultivation. Apart from being uneconomical in the long run (with more inputs), cultivation of such lands, can lead to serious in balances in the eco systems. For such lands, some alternate efficient land use system other than arable cropping would be desirable. In alternate land use system, by encouraging tree and grass components, the demand for food and fodder can be solved

To some extent this stability in production will be achieved along with the safety of environment. As seen from some model watersheds alternate land use systems not only help in generating much needed off season employment but they also help in utilising off season rain which may otherwise go waste. The pasture component in such systems should have the characteristics like growing well even under shade, drought resistance, easy propagation, high palatability, conserving soil and moisture, withstanding over grazing etc. Different legumes like style species and grasses including Cenchrus species have been found performing well in many situations.

Land and water conservation practices

Soil and water conservation practices are the primary step of watershed management program. Conservation practices can be divided into two main categories:

1. *in-situ* and
2. *ex-situ* management.

Land and water conservation practices, those made within agricultural fields like construction of contour bunds, graded bunds, field bunds, terraces building, broad bed and furrow practice and other soil-moisture conservation practices, are known as in-situ management. These practices protect land degradation, improve soil health, and increase soil-moisture availability and groundwater recharge. Moreover, construction of check dam, farm pond, gully control structures, pits excavation across the stream channel is known as *ex-situ* management. *Ex-situ* watershed management practices reduce peak discharge in order to reclaim gully formation and harvest substantial amount of runoff, which increases groundwater recharge and irrigation potential in watersheds.

Components of watershed development

1. Human resource development and community development

Environmental education starts with the creation of awareness about environmental issues (e.g. responsible management of waste) and includes learning about interactions of geo spheres, planting trees, etc. Environmental education can be carried out in schools, universities, families, during community workshops, etc. Environmental education is a process of learning about and for the Watershed Management approach environment so as to work for it and not to destroy it. In the process of *capacity building* regional actors acquire qualification concerning natural resource management and development and organizations develop within the watershed. This is accompanied by processes of change directed to more efficiency and effectiveness as well as process of reform to more democracy, wealth and development. Tools that are used are training and extension.

2. Soil and land management

The basic goal is to conserve natural and cultural areas in a sub watershed in order to sustain the integrity of its ecosystems and maintain desired human uses. Five types of land areas to consider conserving include:

1. critical habitats for plant and animal communities
2. aquatic corridors along streams and shorelines
3. hydrologic reserves to sustain predevelopment flow patterns in water bodies
4. water pollution hazard areas that have potential to pollute water resources
5. cultural/historic areas providing a sense of place

The second basic goal is to reduce sediment loss during construction and to ensure that conservation areas, buffers, and forests are not cleared or otherwise disturbed during construction

1. The period when vegetation is cleared and a site is graded to create a new landscape is the most destructive and it is very difficult to enforce and maintain best management practices.
2. Sediment-loading rates from construction sites are 5 to 500 times greater than those from undeveloped land.

3. Water management

The basic goal is to apply special protection where land and water meet in the form of a buffer. Buffers can be placed along a stream, shoreline, or wetland to:

- filter pollutants (i.e., sediments, nutrients, toxicants, bacteria)
- stabilize soils and reducing channel erosion
- store flood waters
- maintain temperature regime
- maintain habitats
- provide aesthetics and recreational enjoyment

4. Crop management

The crop diversification refers to bringing about a desirable change in the existing cropping patterns towards a more balanced cropping system to reduce the risk of crop failure; and crop intensification is the increasing cropping intensity and production to meet the ever increasing demand for food in a given landscape. Watershed management puts emphasis on crop diversification and intensification through the use of advanced technologies, especially good variety of seeds, balanced fertilizer application and by providing supplemental irrigation.

5. Afforestation, pasture development

In highly degraded areas **reforestation** is an important measure for regeneration of the soil and water balances (see figure). Trees fulfill many functions such as protection from erosion or conserving soil moisture. Two decisions have to be made initially when reforestation is to be done:

- Should the area be completely closed for cultivation and pasture for a specific period? If so, the border must be marked clearly.
- Which works better: direct seeding or planting of nursery-grown seedlings

The answer depends on many factors such as climate, biotic factors, soil, etc.

If a nursery is chosen the appropriate site has to be selected carefully. At least four criteria have to be fulfilled:

- The area of the intended nursery should be near the planting site;
- The area must be large enough (about 0.5 ha for every 100,000 seedlings
- Availability of water year-round;
- The site should have a suitable soil.

The nursery can be managed commercially or by the community. Some know-how concerning soil-mixing ratios, irrigation, selection of species, etc. is required. In general, indigenous species are to be preferred, as experience with exotic species such as eucalyptus has shown in the past. When the seedlings reach an adequate size they are planted out. Later on, when the trees have grown, the forest can also be used by the people, e.g., for bee keeping.

By choosing appropriate soil and water conservation measures in projects it is to be considered that a forestation takes time to show positive impacts.

Agronomic measures are the third type of soil and water conservation measures explained in this Watershed Management-Module.

6. Sustainability and environmental appropriateness

Resources have to be managed in such a way that the demand of the present generation can be satisfied without exhausting, degrading or destroying them in the long term in order to preserve them as a basis of survival for future generations (intergenerational equity). Such a management for the benefit of the people and in harmony with the environment implies that

- Resources are only to be used if they are renewable are used at rates that correspond to the replenishment rates or at an intensity which does not exceed the carrying capacity of the system.

The overall societal development has to be sustainable as well. Sustainability in all its dimensions can only be attained if the goals are defined according to the environmental, socio-economic, political and institutional framework. For example, winter sport tourism cannot be economically and ecologically sustainable in a region only receiving snow five days per year on average. Here has to be prepared by snow guns consuming valuable water and energy resources.

- Pasture/Fodder Development
- Livestock Management
- Rural Energy Management
- Farm and non-farm value addition activities

Farmers those solely dependent on agriculture, hold high uncertainty and risk of failure due to various extreme events, pest and disease attack, and market shocks. Therefore, integration of agriculture (on-farm) and non-agriculture (o/-farm) activities is required at various scales for generating consistent source of income and support for their livelihood. For example, agriculture, livestock production and dairy farming, together can make more resilient and sustainable system compared to adopting agriculture practice alone. Product or by-product of one system could be utilized for other and vice-versa. In this example, biomass production (crop straw) after crop harvesting could be utilized for livestock feeding and manure obtained from livestock could be applied in 1eld to maintain soil fertility. It includes horticulture plantation, aquaculture, and animal husbandry at indivisible farm, household or community scale.

Water only cannot increase crop productivity to its potential level without other interventions. A balanced nutrient diet along with adequate moisture availability and pest and disease free environment can turn agricultural production several folds higher compared to unmanaged land. Integrated nutrient management (INM) involves the integral use of organic manure, crop straw, and other plant and tree biomass material along with little application of chemical fertilizer (both macro and micro-nutrients).

Integrated pest management (IPM) involves use of different crop pest control practices like cultural, biological and chemical methods in a combined and compatible way to suppress pest infestations. Thus, the main goals of INM and IPM are to maintain soil fertility, manage pest and the environment so as to balance costs, bene1ts, public health, and environmental quality.

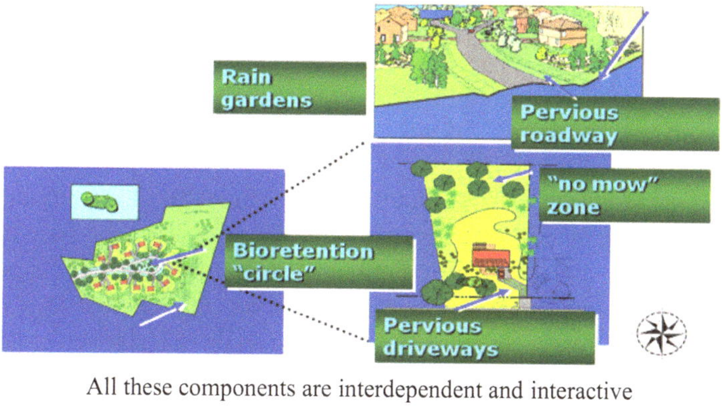

All these components are interdependent and interactive

Steps in implementation of watershed

Effective watershed management includes both planning and implementation components. While the development of a watershed plan is a critical step in the process, the plan must be successfully implemented before results can be seen. The watershed management process uses a series of cooperative, iterative steps to characterize existing conditions, identify and prioritize problems, define management objectives, and develop and implement protection or remediation strategies as necessary implementing the watershed management plan involves using a variety of expertise and skills, including project management, technical expertise, group facilitation, data analysis, communication, and public relations. Your watershed plan implementation team should include members that can bring these skills to the table. The management measures selected, schedules and milestones established, financial and technical resources identified, and I/E programs developed in the course of assembling your plan provide a road map for implementation. Follow it. Take advantage of the partnerships you formed during plan development to work toward efficient implementation of the plan.

Key implementation activities include the following:

- Ensuring technical assistance in the design and installation of management measures
- Providing training and follow-up support to landowners and other responsible parties in operating and maintaining the management measures
- Managing the funding mechanisms and tracking expenditures for each action and for the project as a whole
- Conducting the land treatment and water quality monitoring activities and interpreting and reporting the data
- Measuring progress against schedules and milestones
- Communicating status and results to stakeholders and the public

Definition of watershed

A watershed, also called a drainage basin or catchment area, is defined as an area in which all water sowing into it goes to a common outlet. People and livestock are the integral part of watershed and their activities affect the productive status of watersheds and vice versa. From the hydrological point of view, the different phases of hydrological cycle in a watershed are dependent on the various natural features and human activities. Watershed is not simply the hydrological unit but also socio-political-ecological entity which plays crucial role in determining food, social, and economical security and provides life support services to rural people.

Delineation of watershed

Hydrologically, watershed is an area from which the runoff flows to a common point on the drainage system. Every stream, tributary, or river has an associated watershed, and small watersheds aggregate together to become larger watersheds. Water travels from headwater to the downward location and meets with similar strength of stream, then it forms one order higher stream. The stream order is a measure of the degree of stream branching within a watershed. Each length of stream is indicated by its order (for example, 1rst-order, second- order, etc.). The start or headwaters of a stream, with no other streams flowing into it, is called the 1st order stream. First-order streams flow together to form a second-order stream. Second-order streams flow into a third-order stream and so on. Stream order describes the relative location of the reach in the watershed. Identifying stream order is useful to understand amount of water availability in reach and its quality; and also used as criteria to divide larger watershed into smaller unit. Moreover, criteria for selecting watershed size also depend on the objectives of the development and terrain slope. A large watershed can be managed in plain valley areas or where forest or pasture development is the main objective. In hilly areas or where intensive agriculture development is planned, the size of watershed relatively preferred is small.

Developed concepts and approaches for watershed management

Realizing the importance of upland areas conservation, great efforts were undertaken, all over the world, to reverse watershed degradation and to reduce its effects on land resources management. However, the desired objectives were not achieved in most of the cases as a result of inappropriate concepts and approaches. These concepts and approaches were identified as carrying the following major shortcomings:

- Giving priority to natural resources conservation;
- With little focus on human activities and people priorities and needs;
- Neglecting beneficiaries involvement and contribution in the planning and implementation of the watershed management interventions.

The watershed management recently developed approaches are emphasising on beneficiaries involvement and participation as active partners in making decisions with regards to the planning and implementation of watershed management development programmes. In this respect, it was recognised in recent years that watershed management cannot be only treated through methods and tools of natural sciences. Additional aspects dealing with economic, social, political and cultural exchanges were then considered as required to achieve the watershed management expected results. This implies that watershed management is an inter-disciplinary field which requires the performance of different disciplines and the cooperation and coordination among all concerned.

Components of watershed management

Entry point activity (EPA)

Entry Point Activity is the 1rst formal project intervention which is undertaken after the transect walk, selection and 1nalization of the watershed. It is highly recommended to use knowledge-based entry point activity to build the rapport with the community. Direct cash-based EPA must be avoided as such activities give a wrong signal to the community at the beginning for various interventions. Details of the knowledge-based EPA to build rapport with the community ensuring tangible economic bene1ts to the community members are described here.

Land and water conservation practices

Soil and water conservation practices are the primary step of watershed management program. Conservation practices can be divided into two main categories: 1) *in-situ* and 2) *ex-situ* management. Land and water conservation practices, those made within agricultural 1elds like construction of contour bunds, graded bunds, 1eld bunds, terraces building, broad bed and furrow practice and other soil-moisture conservation practices, are known as in-situ management (Figure 3). These practices protect land degradation, improve soil health, and increase soil-moisture availability and groundwater recharge. Moreover, construction of check dam, farm pond, gully control structures, pits excavation across the stream channel is known as *ex-situ* management (Figure 4). *Ex-situ* watershed management practices reduce peak discharge in order to reclaim gully formation and harvest substantial amount of runo/, which increases groundwater recharge and irrigation potential in watersheds.

Integrated pest and nutrient management

Water only cannot increase crop productivity to its potential level without other interventions. A balanced nutrient diet along with adequate moisture availability and pest and disease free environment can turn agricultural production several folds higher compared to unmanaged land. Integrated nutrient management (INM) involves the integral use of organic manure, crop straw, and other plant and tree biomass material along with little application of chemical fertilizer (both macro and micro-nutrients). Integrated pest management (IPM) involves use of different crop pest control practices like cultural, biological and chemical methods in a combined and compatible way to suppress pest infestations. Thus, the main goals of INM and IPM are to maintain soil fertility, manage pest and the environment so as to balance costs, bene1ts, public health, and environmental quality.

Crop diversification and intensification

The crop diversi1cation refers to bringing about a desirable change in the existing cropping patterns towards a more balanced cropping system to reduce the risk of crop failure; and crop intensi1cation is the increasing cropping intensity and production to meet the ever increasing demand for food in a given landscape. Watershed management puts emphasis on crop diversi1cation and intensi1cation through the use of advanced technologies, especially good variety of seeds, balanced fertilizer application and by providing supplemental irrigation.

1. Use of multiple resources

Farmers those solely dependent on agriculture, hold high uncertainty and risk of failure due to various extreme events, pest and disease attack, and market shocks. Therefore, integration of agriculture (on-farm) and non-agriculture (off-farm) activities is required at various scales for generating consistent source of income and support for their livelihood. For example, agriculture, livestock production and dairy farming, together can make more resilient and sustainable system compared to adopting agriculture practice alone. Product or by-product of one system could be utilized for other and vice-versa. In this example, biomass production (crop straw) after crop harvesting could be utilized for livestock feeding and manure obtained from livestock could be applied in 1eld to maintain soil fertility. It includes horticulture plantation, aquaculture, and animal husbandry at indivisible farm, household or community scale.

2. Capacity building

Watershed development requires multiple interventions that jointly enhance the resource base and livelihoods of the rural people. This requires capacity building of all the stakeholders from farmer to policy makers. Capacity building is a process to strengthen the abilities of people to make effective and efficient use of resources in order to achieve their own goals on a sustained basis. Unawareness and ignorance of the stakeholders about the objectives, approaches, and activities are the reasons that affect the performance of the watersheds. Capacity building program focuses on construction of low cost soil and water conservation methods, production and use of bio-fertilizers and bio-pesticides, income generating activities, livestock based activities, waste land development, market linkage for primary stakeholders. Clear understanding of strategic planning, monitoring and evaluation mechanism and other expertise in 1eld of science and management is essential for government officials and policy makers. The stakeholders should be aware about the importance of various activities, their bene1ts in terms of economics, social and environmental factors. Therefore, organizing various training at different scales is important for watershed development.

Watershed management approaches

Integrated approach

This approach suggest the integration of technologies within the natural boundaries of a drainage area for optimum development of land, water, and plant resources to meet the basic needs of people and animals in a sustainable manner. This approach aims to improve the standard of living of common people by increasing his earning capacity by offering all facilities required for optimum production. In order to achieve its objective, integrated watershed management suggests to adopt land and water conservation practices, water harvesting in ponds and recharging of groundwater for increasing water resources potential and stress on crop diversification, use of improved variety of seeds, integrated nutrient management and integrated pest management practices, etc.

Consortium approach

Consortium approach emphasizes on collective action and community participation including of primary stakeholders, government and non-government organizations, and other institutions. Watershed management requires multidisciplinary skills and competencies. Easy access and timely advice to farmers are important drivers for the observed impressive impacts in the watershed. These lead to enhance awareness of the farmers and their ability to consult with the right people when problems arise. It requires multidisciplinary profeciency in field of engineering, agronomy, forestry, horticulture, animal husbandry, entomology, social science, economics and marketing. It is not always possible to get all the required support and skills-set in one organization. Thus, consortium approach brings together the expertise of different areas to expand the effectiveness of the various watershed initiatives and interventions.

Recommendations

1. Select watershed sites where dire need exists in terms of improving soil and water conservation, enhancing productivity and improving livelihoods.
2. Adopt holistic and participatory consortium approach from the beginning i.e., from selection of watershed.
3. Ensure that ground rules for operation are made clear to the community as well as consortium partners.
4. Adopt knowledge-based entry point approach to build rapport with the community and ensure tangible economic benefits for the community.

Types of watershed

Watershed is classified depending upon the size, drainage, shape and landuse pattern:

1. Macro watershed (> 50,000 ha)
2. Sub-watershed (10,000 to 50,000 ha)
3. Milli-watershed (1000 to10000 ha)
4. Micro watershed (100 to 1000 ha)
5. Mini watershed (1-100 ha)

Watershed management

a) This is the process of guiding &organising land and other resource usage in a watershed

b) Ensuring the sustenance of the environment
   - i) Mainly the soil and water resources
   - ii) need to recognise the interrelationships between, land use, soil-water, and slope of terrain

c) Unifying focus in watershed management is in how various
   - i) human activities affect the relationship between water and
   - ii) other natural resources

d) Provides a basis for actions concerning the development and conservation

Factors affecting watershed management

a) Watershed characters
   - i) Size and shape
   - ii) Topography
   - iii) Soils
   - iv) Relief

b) Climatic characteristic
   - i) Precipitation
   - ii) Amount and intensity of rainfall

c) Watershed operation

d) Land use pattern
   - i) Vegetative cover
   - ii) Density

e) Social status of inhability

f) Water resource and their capabilities

Watershed management practices

1. In terms of purpose
   i) To increase infiltration
   ii) To increase water holding capacity
   iii) To prevent soil erosion
2. Method and accomplishment; In brief various control measures are:
   i) Vegetative measures (Agronomical measures)
   ii) Strip cropping
   iii) Pasture cropping
   iv) Grassland farming
   v) Woodlands
2. Engineering measures (Structural practices)
   i) Contour bunding
   ii) Terracing
   iii) Construction of earthern embankment
   iv) Construction of check dams
   v) Construction of farm ponds
   vi) Construction of diversion
   vii) Gully controlling structure
   viii) Providing vegetative and stone barriers
   ix) Construction of silt tanks

Rainwater harvesting is the main component of watershed management. Some of the watershed management structures are as follows.

Broad beds and furrows

a) Function

To control erosion and to conserve soil moisture in the soil during rainy days.

b) General information

The broad bed and furrow system is laid within the field boundaries. The land levels taken and it is laid using either animal drawn or tractor drawn ridgers.

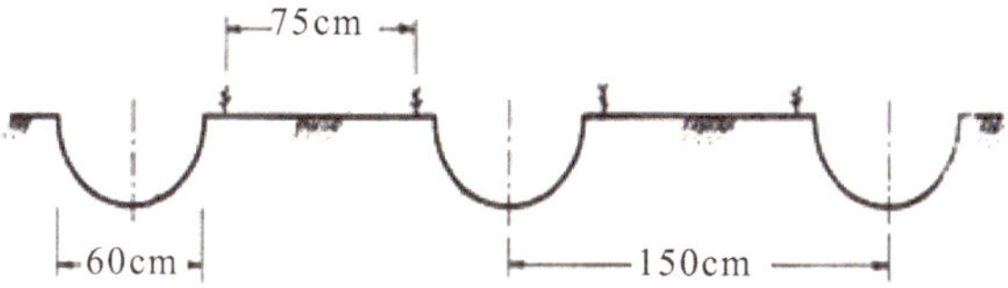

Maize Crop

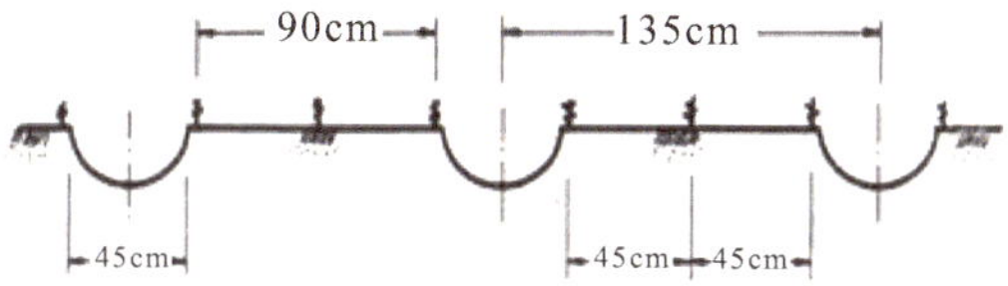

Sorghum Crop

Broad beds and furrows

c) Cost

Approximate cost for laying beds & furrows is Rs.1800 / ha

d) Salient features

i) Conserves soil moisture in dryland

ii) Controls soil erosion

iii) Acts as a drainage channel during heavy rainy days.

2. Contour bund

a) Function

To intercept the run off flowing down the slope by an embankment

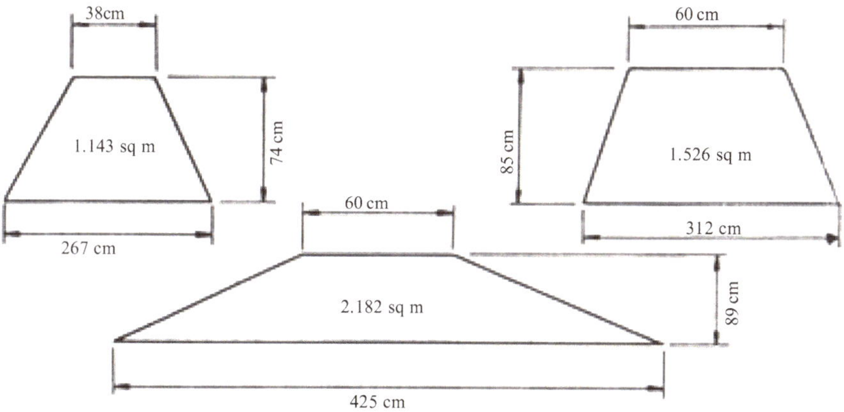

Contour bunding

b) General information

It helps to control run off velocity. The embankment may be closed or open, surplus arrangements are provided wherever necessary.

c) Cost

Approximate cost of laying contour bund is Rs.1400 / ha

d) Salient features

i) It can be adopted on all soils

ii) It can be laid upto 6% slopes.

iii) It helps to retain moisture in the field

3. Bench terracing

a) Function

It helps to bring sloping land into different level strips to enable cultivation

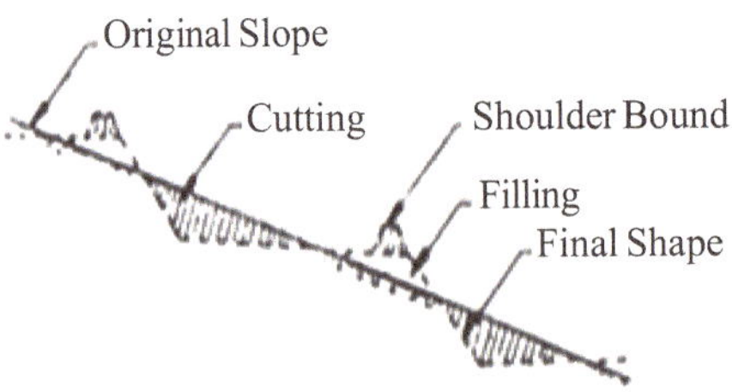

1. Level and table top

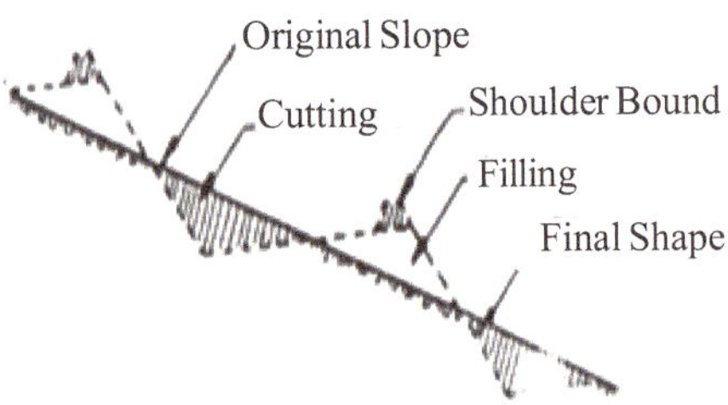

2. Sloping inward

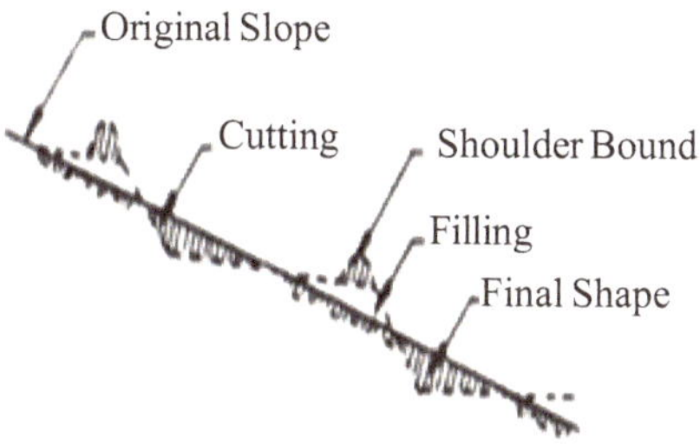

1. Sloping outward

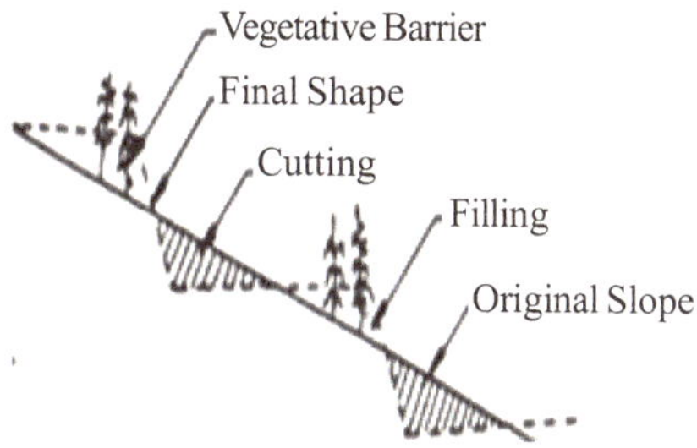

4. Puertorico type terrace

Types of Bench Terraces

b) General information

It consists of construction of step like fields along contours by half cutting and half filling. Original slope is converted into level fields. The vertical & horizontal intervals are decided based on level slope.

c) Cost

Approximate cost for laying the terrace is Rs.5000 / ha

d) Salient features

i) Suitable for hilly regions

ii) The benches may be inward sloping to drain off excess water

iii) The outward sloping benches will help to reduce the existing steep slope to mild one

iv) It is adopted in soils with slopes greater than 6%.

4. Microcatchments for sloping lands

a) Function

It is useful for in-situ moisture conservation and erosion control for tree crops

b) General information

Technique	Storage capacityper unit (m³)	Annual run-off contribution to soil moisture Additional water stored per ha (m³)	Estimated surface Run-off control (%)
Triangular catchments (V-Bunds)	13	5200	80
Crescent bunds	10.2	4080	73

c) Cost

Technique	Cost/ha (Rs.)
Triangular catchments (V-Bunds)	6000-7000
Crescent bunds	2500-3000

d) Salient features

- Slope ranges from 2-8%
- Soil type – Light to moderate texture
- In-situ moisture conservation with staggered planting
- Suitable for dryland Horticulture & Agroforestry
- Bund height – 30 to 45 cm

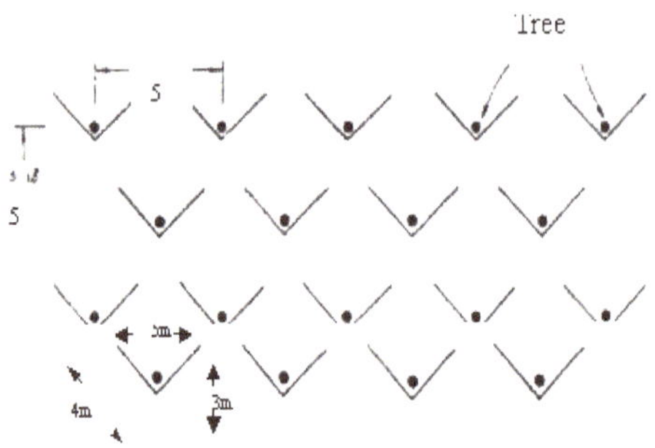

Microcatchments

5. Check dam

- A low weir normally constructed across the gullies
- Constructed on small streams and long gullies formed by erosive activity of flood water

Salient features

- It cuts the velocity and reduces erosive activity
- The stored water improves soil moisture of the adjoining area and allows percolation to recharge the aquifers
- Spacing between the check dams water spread of one should be beyond the water spread of the other
- Height depends on the bank height, varies from a metre to 3 metre and length varies from less than 3m to 10m
- Cost varies from Rs. 40000/- to Rs. 100000/- per unit

6. Percolation pond

Function

To augment the groundwater recharge

Salient features

- Shallow depression created at lower portions in a natural or diverted stream course
- Preferable under gentle sloping stream where narrow valley exists
- Located in soils of permeable nature
- Adaptable where 20-30 ground water wells for irrigation exist within the zone of influence about 800-900m
- Minimum capacity may be around 5000 m^3 for the sake of economy
- Also act as silt detention reservoir
- Cost varies from Rs. 60000 to 150000 per unit

7. Stone barrier

Components of watershed management programme

The main components of watershed management are:

1. Water resource improvement
2. Soil water conservation and water harvesting
3. Crop management
4. Alternate land use systems and integration of livestock in farming system
5. Land treatment in non arable lands

Successful implementation of the above three components in the watersheds would lead to the development of livestock, poultry, pisciculture and other associated and allied activities, finally resulting in improved standard of living of the farmers and rural masses.

Water resource improvement

Various water resources should be improved in the watershed command areas. The water from these resources should be collected in the watershed which can be used for the crop at critical growth stages.

Table: Agro-ecological conditions in different rainfall zones of India

Mean annual rainfall (mm)	Predominant soil types	Mean annual temperature (^{0}C)	Length of growing season under rainfed condition (days)	Predominant rainfed crops
< 400	Sandy	20-30	30-80	Grasses, Pearlmillet
		0-10	30-90	Short duration pulses
400-1000	Sandy soils (Aridisols); Red soils (Alfisols and relates soils);	20-30	80-200	Pearlmillet, sorghum maize, oilseeds, pulses
	Black soils (vertisols and vertic inceptisols)	-	-	Cotton
1000-1800	Alluvial soils (Entisols); Laterite soils (Alfisols and relates soils); Balck soils (vertic and vertic inceptisols)	20-30	200-300	Maize, Paddy, wheat, Barley, Mustard and Soybean
> 1800	Sub-montane soils (inceptisols)	15-20	> 300	Paddy, plantation crops

Depending on the water availability periods, examples of some selected crops for rainy season and post rainy season are given in Table below:

Table: Major efficient food crop-based double cropping systems for different rain dependent regions of India (based on water availability periods)

Water availability period (days)	Rainy season crops	Post rainy season crops
110-150	Cowpea/blackgram	Safflower/seasons crops
	Soybean	Mustard/safflower
	Greengram	Chickpea/barley
150-175	Greengram	Sorghum
175-200	Cowpea	Sorghum
	Greengram	Safflower
	Balckgram	Barley/mustard
	Pearlmillet	Chickpea
	Maize	Wheat/wheat+
		Chickpea/mustard
	Rice	Chickpea
	Sesame	Chickpea
200-250	Sorghum/groundnut/maize	Safflower
	Soybean	Wheat
		Safflower/chickpea
		Soybean/maize
> 250	Rice/maize/finger millet/ groundnut/pearl millet	Wheat/chickpea/linseed/lentil/ horsegram/barley
	Soybean	Fingermillet
	Pearlmillet	Wheat

Many drylands of the country are not only thirsty, these are hungry as well. Optimum doses of fertilizers are, therefore, recommended for the crops depending upon physico chemical properties of soil. And farmers are advised for proper crop residue managementstubbles of the plants being left in the field for decomposition and building up the fertility. However, major breakthrough has yet to be achieved in cropping systems in the watersheds by changing over to organic farming with treebased cropping programmes. The research work is being initiated on these lines through alternate land use systems.

Alternate landuse systems and integration of livestock in farming system

In the watersheds, most of the uplands are degraded and these are also marginal lands with low productivity level. Due to the population pressure, more and more marginal and sub marginal lands are being brought under cultivation. Apart from being uneconomical in the long run (with more inputs), cultivation of such lands, can lead to serious in balances in the eco systems. For such lands, some alternate efficient land use system other than arable cropping would be desirable. In alternate land use system, by encouraging tree and grass components, the demand for food and fodder can be solved to some extent. Thus, stability in production will be achieved along with the safety of environment. As seen from some model watersheds alternate land use systems not only help in generating much needed off season employment but they also help in utilizing off season rain which may otherwise go waste.

In general, land capability classes are the determinant factors for deciding different alternate land use options. By and large, in land classes I to Ill, arable farming can be practiced and beyond that lands should be utilized for forestry and grassland management ut from economic considerations, agro forestry potions integrating farming with other land uses like horticulture, agriculture and pasture may find acceptance from the farmers. The evaluation of all these systems is under research investigations. The tree component of the system should be selected according to the soil and climatic conditions. The trees should preferably have the characteristics like fast growing with high palatability, good coppicing ability, and drought resistance also having properties like adding organic manure in soil profile. The pasture component in such systems should have the characteristics like growing well even under shade, drought resistance, easy propagation, high palatability, conserving soil and moisture, withstanding over grazing etc. Different legumes like style species and grasses including Cenchrus species have been found performing well in many situations.

Dryland horticulture is promising enterprise in slightly favorable soil conditions especially where there is provision of water harvesting. In this hardy fruit species such as ber, pomegranate, custard apple, guava and aonla may have to be preferred over others.

Many drylands of the country are not only thirsty, these are hungry as well. Optimum doses of fertilizers are, therefore, recommended for the crops depending upon physico chemical properties of soiland farmers are advised for proper crop residue management and stubbles of the plants being left in the field for decomposition and building up the fertility. However, major breakthrough has yet to be achieved in cropping systems in the watersheds by changing over to organic farming with tree based cropping programmes. The research work is being initiated on these lines through alternate land use systems.

Integrated farming system having the integration of crops with livestock, poultry, fish, mushroom, bee keeping etc. should be practiced keeping the resources in view. Different models of farming systems suited to the farming conditions of the individual should be adopted.

Chapter 10

Watershed Management in India

Management of watershed in India

It involves management of water, land, energy and greenery, integrating all the relevant scientific approaches appropriate to socio-economic background for a pragmatic development of a watershed. In practice, it means allowing load free rainfall run off from the watershed. Clear outflow imply management of effective soil and water conservation and generating a lush green carpet in whole of the area. In other worlds, it is greening of a watershed through proper management of land, water and energy sources. The comprehensive development of watershed seems to make productive use of all its natural resources and also protect them is termed as watershed management.

The Banave terraces in Indonesia and those in Nepal, Mediterranean are examples of integration of human elements into a system of land management, which is an early achievement in watershed management. Watershed management is not merely anti-erosional and anti-run off approach but also a comprehensive integrated approach of land and water resources management. The approach is preventive, progressive, corrective as well as curative. Here wasted land resources could be fruit fully harvested into a productive face for rural development.

The chief objectives of watershed management are:

- Conserving soil and water
- Improving the ability of land to hold water
- Rain water harvesting and recharging
- Growing greenery trees, crops and grasses

In order to obtain maximum benefit from technological development, it is imperative that the natural resources, soil and water be properly protected and judicially utilized to improve their productivity constantly. When a watershed is properly managed for water, it is also property managed for soil and vegetation. Watershed being natural hydrological entities, they respond effectively to various engineering, biological and cultural treatments designed to maximize production.

The different management practices are:

1. Hydrological

Various analysis for different agro-climatic zones and sub-zones of the country have shown clearly that overwhelming majority of the production areas (including some in sub-humid and humid regions) is subject to water stress and drought conditions. This is not only due to low annual rainfall but also erratic distribution during the year and over the years apart from unfavorable rains and improper land use system. Realizing this, hydrological manipulations through cover management and structural (micro) measures had received priority attention.

No plant can survive / grow without a minimum quality of water and of appropriate quality. This, the conservation, augmentation and regulation of the supply of moisture vis-à-vis various plant growth stages been recognized as the most important aspect in the redevelopment programme. Since the supply of rainfall is not under the control of human beings, only manipulation of the incident rainfall or ground water could help achieve this objective to a great extent. Again all users, such as plant, animal and man can get water after it has routed through land surface and soil profile. Thus *in situ* moisture conservation techniques and *in-situ* and near site water harvesting system should be the pre-requisites in water shed management plans.

a) *In situ* moisture conservation and H_2O harvesting system

Managing and making available moisture to the primary production systems as well as the plants which are used for stabilization of protection of degraded/ degrading areas, is pre-requisite. For *in situ* conservation over extensive areas, mechanical barriers across the slope have in use both on arable as well as non-arable areas. These are effective to reduce velocity and erosive power of run off and also promote absorption specially in immediate upstream and down stream area. However, for uniform and good performance of plants in terms of regeneration and growth, uniform absorption and distribution of moisture over as large as possible is needed. This is calls for micro treatment such as dead and contour furrows, racking etc. in between the barriers across the slope. Practice of growing grasses and legumes in the intervening area goes a long way to distribute the benefit of conservation and augmentation of moisture over extensive area.

Complete replacement of mechanical barriers by vegetative ones may be possible only at some sites where slope is not very steep, over land run off is not a very big problem and process of erosion is not very advance. More important and technically sound measure is to develop the net work of micro water harvesting structure over extensive areas. This would directly enhance the possibility of quicker establishment of plants and their better growth and development. There is a scope to improve design specification and lay out of such systems specially by adopting some of the details from the traditional system such as nullah ploughing, nullah bunding, bunchis, small ponds and other models in use for generations in arid, semi-arid and even in sub-humid areas.

b) Cover management

The moisture conservation and augmentation would not be lasting by themselves. All these measures need to be taken up with the complementary support of vegetative regeneration. In many areas, the pressure of livestock and small and marginal farmers and landless labourers are great. These areas must yield in short term and number of produce such as fodder, ,fuel, wood and small items for consumption by promoting appropriate agro-forestry models with distinct thrust on improving hydrologic regimes. The relation between contributing areas to storage and submergence area of micro structure need to be critically studied and appropriate guidelines developed for varying site conditions. Similarly interrelationship between volume and recharge and area of rense call for better analysis, for identification of indicators that will help developing design specification. For selection of plant species, greater understanding of the root-system and the contribution to expansion of soil volume that increases watershed retention needs to be studied and appropriate norms should be provided.

2. Biological

Initial thrust of biological measures has been to increase the vegetative cover of varied quantity. But in mid seventies stress on trees seem to have edged out other plant species such as bash, shrub, grass and other legumes. In degraded areas, establishing and promoting tree growth at a faster rate which can protect the land is difficult atleast for a period of two years. Hence the emphasis should be equally on grass and legumes so that a ground cover and quick incorporation of organic residues, including additional nitrogen could be achieved. Emphasis showed be on growing legumes grasses, which not only enrich soil but also offer good fodder which is extremely in short supply all over the country.

In respect of biological measures, mono-planting needs to be replaced by mixed plantation. For selection of species their impact on hydrologic regime should receive more scientific consideration. For degraded lands emphasis should be given to grasses legumes of reduce no. of trees ha^{-1} substantially. The strategy to achieve maximum divident would include different innovative models of agro forestry. According to the need of the area to model should include livestock management, duckeny, viniculture and sericulture to generate economic activities besides small scale enterprises based on biomass.

3. Bio-inputs

Reasonable research evidences are available to open a new dimension in accelerating the pace of establishment and growth of vegetation in degraded areas particularly arid lands, mixed areas, coasts and those affected by industrial and urban wastes, the application of bio-inputs such as mycorrhizae, rhizobiun, and organic chemicals like bactrin, natrin, phospin, etc. to overcome various limitations and increase tolerance of the plants to water stress, chemical imbalances etc. In the watershed management programme, application of these new bio-inputs has yet be initiated and popularized.

4. Waste reuse

The sewage and garbage from the urban settlements and industrial affluent are being disposed through dumping on the low lying areas or wet lands and discharging into water bodies as well as agricultural and other lands. These have been not only polluting water and degrading environment but also rendering large areas unproductive. In the other hand, the nutrients and the fresh H_2O available in these waste products have been successfully used by many in greening and also growing many consumable items. As there are contamination hazards, watershed management programme may atleast initiates utilizing waste H_2O and garbage as a source of nutrient and H_2O for redeveloping the degraded areas which such risks don't exist or can be utilized.

5. Land management

Land management is another very crucial factor in watershed management. Proper land management decreases the runoff loss, increases H_2O recharge and keeps the soil fertile. The management practices are:

i) Ploughing: Plough and ploughing are part and pareel of village life. Ploughing along the slope increases unoff, loses 50-80% of rainfall, decreases soil moisture and finally erodes soil. Just ploughing along the contains, soil erosion can be reduced upto 50 per cent.

ii) Furrowing: Ploughing deeper for even 2-10 lines is another measure for consumption. This practice coupled with intercropping and hedging give good results. Ploughing leaves minor furrows and they harvest rainfwater flows. This process is further accelerated by domble ploughing along the row for a chosen spacing depending upon the scope. If hedging and planting are carried out, this practice contains > 50 rainfall.

   A broad bed and furrow system (BBF), involving graded, wide beds separated by furrow draining into grassed water ways in experimented by ICRISAT with good results. The BBF system is laid out on grades 0.4-0.8% for optimum performance. The raised position acts as bund ensuring soil stability while the shallow furrow provides change and erosion. This system is flexible with variables spacing for accommodating different crops, continuous furrowing along the contours and ridge development across the furrows is another practice adopted in low rainfall regions for cent percent harvesting.

iii) Trenching : Narrow excavation along the contours is another conservation technique. The economical dimensions are 0.3 m width and 0.6 m depth in lands with low slopes or good rains. While the dimensions can be interchanged in genetly sloping fields or when low intensity rains are common. Whereas the degree of erosion is very high due to soft soil, intensive rains or high run off and velocity, higher dimension of the trenching are to be adopted. This soil excavated is to be conveniently but as a bend on the down slope side. Trenching proves to be a good conservation technique, especially when plants are grown along the trench for utilizing the high moisture available along the trench.

iv) Bunding : Bunding along the contours controls the erosion very effectively. In well drained red soils in low rainfall areas, bunding is an accepted practice. The existing stream system is utilized as waterways. Furrow bunding is the only option in certain sleep sloping lands. There are 3 types of bunding, graded, inter and compartment.

Graded bunding consists of bunding with 0.2% gradient for checking he overland flow. For accommodating storm rain water, water way system is developed using the existing streams. For efficient soil conservation gully stabilization measures are a also carried out. In interbunding, gaps are left alternatively for conserving more rainfall through reversing the direction of water spreads in between the beds. In this method, the computation must be precise. In compartmental bunding the lands are bunded perpendicular to contour also for cent per cent harvesting of the rainfall in semi-arid and desert regimes. It is an excellent practice for growing greenery in low rainfall regions.

v) Gradoni : Narrow trenches built along contours for collecting overland flow and increasing soil moisture are known as gradoni. They are either continuous, intermittent or build for individual plants. In slopes upto 30% level trenches are built; on steep slopes of 30-70% reverse sloped trenches are practiced; while for dissected slopes with undulating terrain short or individual gradoni are prepared. For uniform slopes, continuous gradoni, synonemous to minor trenching are prepared.

vi) Hedging : The method not only solves the conservation problem, but also produces biomass and stablizes the ground further by root system. No additional costs are involved. Though vetiver grass is strongly recommended, subabul hedging produces good biomass for feeding the cattle with nutritious feed. Hedging gives excellent results in catching rain water and improving soil moisture. It also forms a wind breaker contains even storm winds as system on regional scale. Hedging is practiced from gentle to moderate sloping lands for increasing crop yields. Hedges are grown either the furrows, trenches or an bunds. They are grown across the stream for conserving flood load. When practiced in combination with other methods, they conserve almost all the rain water in drought prone areas.

vii) Terracing : Bunding and terracing are common practice among the Jamaican high lands. The effectiveness of terraces depends also on green cover management. Bench terraces are a series of uniform, continous, level strips running across the slope at vertical intervals supported by steep banks. The banks are built by earth and protected by grass or easily accessible racks. They are good for irrigation utilizing the rain water and interflow between the rains. In high rainfall regions outward slopes benches allow free flow or water without benching the terraces. In low rainfall region reverse scope terraces help in water harvesting. These two types of terraces are used for crops mainly. One advantage of sloped terraces is that the cost of banks or raisor is reduced. Intermittent terracing is another combination of level benching with outward sloping terraces. In hill side ditches, level benching and onward sloping berth terraces are combined. This practice is good for semi-permanent rainfed crops. Individual bains and outward terraces are narrow benches alternating with-intward slopes. Fruit plants can be implanted either in pits or basin depending upon the soil profile and water requirement. Hexagon pattern is practiced in this system in heaving rainfall regions for crops requiring good transport routes. In convertible terraces, bench terraces sand witching the basin. This practice is ideal for mixed farming. The slopes of all these terraces are protected by grasses or crops.

viii) Measures against shifting cultivation : Agro-horti-pastoral system is a good alternative the shifting cultivation. Here top sleep hills are used for forest trees followed by horticulture and the lower portion is used for agriculture. Local resources based soil and water conservation measures like contour bund, bench terraces, half moon terraces and grass ways are adopted. Watershed based farming system thus ensures adequate protection of land against soil erosion, retain maximum rain water without affecting the crop and the harvested water can be rentilised for pisciculture or any other purpose according to local used.

Impact of watershed programme on sustainable agriculture

The input of this programme has been beyond expectations. The watershed management programme now in operation in different parts of the country has achieved surprising degree of success. The programme are (1) watershed approach for soil and H_2O conservation (2) DPAP, (3) DF, (3) river valley projects, (5) flood control and (6) hill area development.

At places where watershed have been in operation:

1. Soil and H_2O run off has been checked
2. Denuded forests and devastated pasture lands have been rejuvenated
3. Barren and waste lands have been brought under cultivation
4. Groundwater level and
5. Crop yields have gone up and so have the
6. Farmer's income.

The impact and effectiveness of the watershed initiatives such as Sukhomagri and Bunga have clearly shown that it is possible embed ecologically sound rainfed agriculture in economically sustainable development. The other good examples are made at Mir (Udhampur, J & K); Algai Pandiyapuram (Tirunelveli, T.N.), Nartora (Raipur, M.P.), Jhanwar (Jodhpur), Gursutinala (Palamau, Bihar), Padal Singhi (Beed, Maharastra and Dapoli (Mahrastra). The success of Bunga watershed in Yamuna Nagar Haryana is another good example. As a result of water resources development by construction of an earthen dam and associated soil consemation and efficient use of irrigation water and dry farming practices besides improvement in animal health the total food grain production increased from 20 tons to 151 tons and fodder from 56 to 1095 t/year and milk from 1231 to 3406 litre day. The entire cost (36 lacs) of the project was recovered in 3 years and subsequently the watershed society could generate about 1.0 lacs rupee/year.

Chapter 11

Integrated Watershed Development

Integrated watershed management programme (IWMP)

The main objectives of the IWMP are to restore the ecological balance by harnessing, conserving and developing degraded natural resources such as soil, vegetative cover and water. The outcomes are prevention of soil run-off, regeneration of natural vegetation, rain water harvesting and recharging of the ground water table. This enables multi-cropping and the introduction of diverse agro-based activities, which help to provide sustainable livelihoods to the people residing in the watershed area. In addition, there is a Scheme of Technology Development, Extension and Training (TDET) is also being implemented to promote development of cost effective and proven technologies to support watershed management.

Till 1.4.2008, Department implemented 3 watershed programmes viz. Integrated Wastelands Development Programme, Drought Prone Areas Programme, Desert Development Programme. Since then, they have been brought under a comprehensive programme named Integrated Watershed Management Programme (IWMP) to be implemented under Common Guidelines on Watershed Development, 2008.

District watershed development unit (DWDU)

In districts, where the area under the watershed development projects is about 25,000 hectare, a separate dedicated unit, called the District Watershed Development Unit (DWDU) will be established at the district level, which will oversee the implementation of watershed programme in each district and will have separate independent accounts for this purpose. Where the area under Watershed Development Projects is less than approximately 25,000 hectare, the projects will be implemented in accordance with the existing arrangements. However, in such cases one officer shall be exclusively appointed within the DRDA either on contract or on deputation to coordinate watershed projects at the district level. DWDU will function in close co-ordination with the District Planning Committee. There will also be a representation in DWDU for NREGA, BRGF implementing agencies at the district level. Alternatively, the mechanism of approval and implementation of project by the District Level Committee / collector may continue to prevail.

Project implementing agency (PIA)

The SLNA would evolve appropriate mechanisms for selecting and approving the PIAs, who would be responsible for implementation of watershed projects in different districts. These PIAs may include relevant line departments, autonomous organizations under State/ Central Governments, Government, institutes, research bodies, Intermediate Panchayats, Voluntary Organizations (VOs). However, the following criteria may be observed in the selection of these PIAs:

1. They should preferably have prior experience in watershed related aspects or management of watershed development projects.
2. They should be prepared to constitute dedicated Watershed Development Teams.

Voluntary organizations

Voluntary Organizations (VOs) will have an important role in the programme and their services will be utilized substantively in the areas of awareness generation, capacity building, IEC and social audit among others. As far as direct implementation of the programme is concerned, Voluntary Organizations (VOs) with established credentials may be chosen as PIAs on the basis of detailed criteria as enumerated below.

The Voluntary Organizations (VOs) would need to satisfy the following criteria to be selected as PIA:

a) Should be a registered legal entity of at least 5 years standing.
b) Should have had at least 3 years of field experience in the area of community based Natural Resource Management and livelihood development.

Watershed development team

The WDT is an integral part of the PIA and will be set up by the PIA. Each WDT should have at least four members, broadly with knowledge and experience in agriculture, soil science, water management, social mobilization and institutional building. At least one of the WDT members should be a woman. The WDT members should preferably have a professional degree. However, the qualification can be relaxed by the DWDU with the approval of SLNA in deserving cases keeping in view the practical field experience of the candidate. The WDT should be located as close as possible to the watershed project. At the same time, it must be ensured that the WDT should function in close collaboration with the team of experts at the district and state level. The expenses towards the salaries of the WDT members shall be charged from the administrative support to the PIA. DWDU will facilitate the training of the WDT members.

Self help groups

The Watershed Committee shall constitute SHGs in the watershed area with the help of WDT from amongst poor, small and marginal farmer households, landless/asset less poor agricultural labourers, women, shepherds and SC/ST persons. These Groups shall be homogenous groups having common identity and interest who are dependent on the watershed area for their livelihood. Each Self Help Group will be provided with a revolving fund of an amount to be decided by the Nodal Ministry.

User groups

The Watershed Committee (WC) shall also constitute User Groups in the watershed area with the help of WDT. These shall be homogenous groups of persons most affected by each work/ activity and shall include those having land holdings within the watershed areas. Each User Group shall consist of those who are likely to derive direct benefits from a particular watershed work or activity. The Watershed Committee (WC) with the help of the WDT shall facilitate resource-use agreements among the User Groups based on the principles of equity and sustainability. These agreements must be worked out before the concerned work is undertaken. It must be regarded as a pre-condition for that activity. The User Groups will be responsible for the operation and maintenance of all the assets created under the project in close collaboration with the Gram Panchayat and the Gram Sabha.

Watershed committee (WC)

The Gram Sabha will constitute the Watershed Committee (WC) to implement the Watershed project with the technical support of the WDT in the village. The Watershed Committee (WC) has to be registered under the Society Registration Act, 1860. The Gram Sabha may elect/appoint any suitable person from the village as the Chairman of Watershed Committee. The secretary of the Watershed Committee (WC) will be a paid functionary of the Watershed Committee (WC). The Watershed Committee (WC) will comprise of at least 10 members, half of the members shall be representatives of SHGs and User Groups, SC/ST community, women and landless persons in the village. One member of the

WDT shall also be represented in the Watershed Committee (WC). Where the Panchayat covers more than one village, they would constitute a separate subcommittee for each village to manage the watershed development project in the concerned village. Where a watershed project covers more than one Gram Panchayat, separate committees will be constituted for each Gram Panchayat. The Watershed Committee (WC) would be provided with an independent rented office accommodation.

Each Watershed Development Project is expected to achieve the following results by the end of the project period

1. All the works/activities that are planned for the treatment and development of the drainage lines, arable and non-arable lands in the watershed area are completed with the active participation and contribution of the user groups and the community at large.
2. The user groups/panchayats have willingly taken over the operation and maintenance of the assets created and made suitable administrative and financial arrangements for their maintenance and further development.
3. All the members of the Watershed Committee and staff such as Watershed Secretary and Volunteers have been given orientation and training to improve their knowledge and upgrade technical/management and community organisational skills to a level that is appropriate for the successful discharge of their responsibilities on withdrawal of the Watershed Development Team from the Project.
4. The village community would have been organised into several, homogeneous self-help groups for savings and other income generation activities which would have achieved sufficient commitment from their members and built up financial resources to be self sustaining.
5. The increase in cropping intensity and agricultural productivity reflecting in overall increase in agriculture production.
6. Increase in income of farmers/ landless labourers in the project area.
7. Increase in groundwater table due to enhanced recharge by watershed interventions.

Suggested Readings

Aamodt, O.S. and Johnston W.A., 1936. Can.J.Res.14: 122-152.

Adams, J.E., 1962. Effect of soil temperature on grain sorghum growth yield Agron. J. 54: 257-261.

Chen, D.; Sarid, A. and Katchalski, E., 1968. Proc. Nat.Acad.Sci; US. 61: 1378-83.

De, G.C., 1989. Fundamentals of Agronomy. Oxford and IBH Pulishing Co. Pvt. Ltd. New Delhi.

Dhruvanarayana, V.V., Sastry, G. and Patnaik, U.S., 1990. Watershed Management.

Dhruvanarayana, V.V., Singh, R.P., Bharadwaj, S.P., Sharma, S.N. Sikka, A.K. and Vittal K.P.R., 1987. Watershed management for drought mitigation. Indian Council of Agricultural Research, New Delhi, p. 38.

El Sharkawy, M. Loomsi, R.S. and Williams, W.A., 1968. J. Applied Ecol. 5: 241-51.

El Sharkawy, M and Hesketh, 1965. Crop Sci. 5: 517-21.

Eveneari, M., 1962. Arid Zone Research, 18: 175-196.

Fuehring, H.D., 1973. Agron. J. 65: 348-51.

Gautam, R.C., 1979. Madras Agric. J. 66: 230-233.

Gautam, R.C., 1990. Proc. Int. Symp. On NRM, N.Delhi, 6-10 Feb, 1990.

Gautam, R.C. and Kaushik, S.K., 1987. Annual Report 1987, Division of Agronomy, IARI, N. Delhi.

Gautam, R.C. and Kaushik, S.K., 1988. Indian J. Agric. Sci. 53: 223-224.

Giri, G.., Singh, R.R. and De, R. 1983. Indian J. Agric. Sci., 53: 899-994.

Grundbacher, F.J., 1963. Bot. Rev. 29: 366-81.

Harrold, L.L. and Edwards, W.M., 1959. A severe rainstorm of no tillage corn. 1. Soil Water Conserv. 27: 30.F

Hatch, M.D. and Slack, C.R., 1970. Ann. Rev. Pl. Physiol. 21: 141-162.

Hatch, M.D., Slack, C.R. and Johsnon, H.S., 1967. Biochem. J. 102: 417-22.

Hossain, M.A. and Maniruzzaman, A.F.M., 1992. Response of rainfed wheat (Triticum aestiuum) to method of sowing placement. Indian J. Agron. 37(4): 710 715.

http://www.downtoearth.org.in/coverage/india-must-focus-on-rainfed-farming-6259)

https://www.gktoday.in/rainfed-agriculture-in-india/) I CAR, New Delhi.

Ibid

ICAR, 1981. The Indo-UK dryland farming project. A five year report, 1975-79. Vol. 1 & 2.

ICRISAT (International Crops Reserach Institute for the Semi Arid Tropics). 1983.Annual Report 1982, 262-264.

Kanitkar, N.V.; Sirur, S.S. and Ghokhale, D.H., 1968. Dry farming in India, ICAR, New Delhi.

Kassas, M.,1966. In Arid lands: A Geographical Appraisal Hills, E. S. (ed), UNESCO, Paris, pp. 145-80.

Kaushik, S.K. and Gautam, R.C., 1984. Indian J.Agric. Sci. 59: 577-580.

Killian, C. and Lemee, G., 1956. In Handbuch der Pfanzenphysiologie, Ruhland, W. (ed), Springer Verlag, Berlin, Band III. Pp. 607-726.

Kramar, P.J., 1959. In Plant physiology Vol.II. Steward, F.C. (ed), Academic press, New York. Pp. 607-726.

Levitt, J., 1958. Protoplasmatologia, 6: 87.

Levitt, J., 1972. Response of plants to environmental stress (ed) Academic press, New York.

Loomis, R.S., Williams, W.A. and Hall, A.E., 1971. Ann. Rev.Pl. Physiol. 22: 431-68.

McDonough, W.T. and Gauch, H.G., 1959. Bull. Md. Agric. Exp. Sta. Pp.103.

Murata, Y. and Iyama, J., 1963. Proc. Crop Sci. Soc. Japan, 31: 315-22.

Oppenheimer, H.R., 1960. Arid zone res. (UNESCO, Paris) 15: 105-38.

Palaniappan, S.P., 1985 Cropping systems in the Tropics. Principles and Management. Wily Eartem Ltd. New Delhi.

Patel, R.R. and Maliwal, G.L., 1982. In: A decade of research achievements and practical recommendations in rainfed farming (1971-1989), GAU, Dantiwara (Gujrat).

Polunin, N., 1960. Introduction to plant Geography and some related sciences, Longmans Green, London and Barnes and Noble, New York.

"PRS Legislative Research" [http://www.prsindia.org/parliamenttrack/vital-stats/status-of-water-availability-for-agriculture-in-india-4339/]

Qureshi, S., 1989. Techniques of dryfarming, In: Regional prespective on dryfarming, pp. 79-95.

Reddy, T.Y. and Reddi, G.H.S., 1994. Principles of Agronomy, pp. 291-292.

Reddy, T.Y. and Reddy, G.H., 2002. Principles of Agronomy. Kalyani Publisher, New Delhi, India, pp 368 407.

Salim, M.H., Todd, G.W. and Schlehuber, A.M., 1965. Ibid. 57: 603-7.

Sanghi, H.K., 1985. Tillage operations in maize, In: Efficient management of dryland crops, CRIDA Hydrabad, pp. 67.

Sharda, U.N., Sharma, N.K., Mohan, S.C. and Khybri, M.L., 1999b. Green manuring for conservation and production in western Himalayas: 2. Effect on moisture conservation, weed control and crop yield. Indian J. Soil Cons. 27(1): 31 35.

Sharda, V.N., Khybri, M.L., Sharma, N.K., Mohan, S.C. and Goyal, G.P., 1999a. Green manuring for conservation and conservation and production in western Himalayas. Indian J. Soil Cons. 27(1): 26 30.

Shields, L.M., 1958. In Morphology in relation to xerophytism (ed) New Mexico Highlands Univ. Bull. Pp. 15-22.

Shimshi, D. and Ephrat, Y., 1970. Report to Ford Foundation.

Siddiqi, K.A., 1987. Contribution of Dactyloclenium aegyptium (L.) Beauv. to bio reclamation of salt affected soil. Annals of Arid Zone, 26: 301-303.

Singh, A.K., Ashok Kumar, Katiyar, V.S., Singh, K.D.and S.U., 1997. Soil and water conservation measures in semi arid region of South eastern Rajasthan. Indian J. Soil Convs. 25(3): 186 189.

Singh, R.P., 1995. Sustainable Development in Dry land Agriculture in India.

Singh, R.P., 1995. Sustainable Development of dryland Agriculture in India. Scientific Pub., Jodhpur.

Singh, R.P. and Sharma, S., 1989. Managing rainwater to better farmers' lot. Indianfarmin, 39(9): 61-63.

Singh, R.P., 1985. Peall millet. In : Management of Dryland Crops., (Eds. V. Balasubramanian and J. Venkateswarlu), Central Research Institute for Dryland Agriculture, Hyderabad, A.P., pp. 51-62.

Singh, R.P., Sharma, S., Padmanabhan, M.V., Das, SK and Mishra, P.K., 1990. Field Manual on Watershed Management. CRIDA, Hyderabad.

Singh, R.P.; S.K.; Bhaskara Rao and Reddy, M.N., 1990. Towards Sustainable Dryland Agriculture Practices. CRIDA, Hydrabad.

Singh, S.P. and Rao, U.M.B., 1998. Rain water management for stabilishing productivity of drylands,. Indian 1. Drylands Agri. Res. & Dev., 3(3): 203-214.

Singh, S.S., 1997. Recent advances in dryland agriculture. Hand book of Agriculture. Pp. 683-695.

Singh, S.S., 1996. Crop management: Under irrigated and rainfed conditions. Kalyani Pub., Ludhiana.

Srivastava, O.S., 1987. Policy issues of Agricultural Economics. Allied Publishers, New Delhi.

Stalefelt, M.G., 1956. In Handbuch der Pfanzenphysiologie, Ruhland, W. (ed), Springer Verlag, Berlin, Band III. Pp. 351-426.

Steiner, J.L. Day, J.C., Pupendick, R.I., Meyer, R.E. and Bertrand., A.R., 1988. Importing and Sustaining Productivity in dryland regions of Developing countries Advances in Soil Science 8: 79-122.

Stocker, O., 1960. Res. Rev. 15: 63-104.

Stutte, C.A. and Todd, G.W., 1967. Phyton 24: 67-75.

Subramanian, V.B. and Singh, R.P., 1983. Dryland Agriculture Research in India. Trust in the Eightes All India Coordinated Research Project on Dry land Agriculture, Hydrabad.

Tripathi, B., Rao, A.N. and Ramakrishna, A., 1992. Weed Management in dryland crops and cropping system. In : Dryland Agriculture in India State of Art of Research in India. (Ed. L.L., Somani), Scientific Publishers, Jodhpur, pp. 185-210.

Vankaterwarlu, J., 1982. Dryland Agriculture Problems and prospects. A decade of Dryland Agricultural Research in India 1971-80. AICRPDA, Hyderabad.

Venkatesarlu, J., 1983. In: Contingent crop production strategy in rainfed areas under different weather conditions, AICRPDA, Hyderabad, ICAR, pp. 5-12.

Venkateswarlu, J., Vishnumurthy, T. and Padamanabhan, M.V., 1991. Contingent crop production strategy in rainfed area under different weather conditions. Project Bulletin. No. 4. CRIDA.

Verma, B., Chinnamani, S., Bhola, S.N., Rao, D.H., Parsad, S.N. and Parkash, C., 1986. Twenty-five years Research on Soil Water and conservation research in ravines of Rajasthan CSWCRTI, Research Center, Kota, 1: 2-15.

Verma, B., Prasad, S.N., Parkash, C., Singh, Ratan and Rao, D.H., 1990. Runoff and soil loss under sorghum at one per cent slope in Kota clay soil. Indian J. Soil Cons., 18(1 & 2): 15-19.

Vijayalakshmi, K., 1987. Rain water management in drylands. In Technological Advances in Dryland Agriculture, CRIDA, Hyderabad, pp. 19-45.

Went, F.W., 1965. In: Desert Research, Research Council of Isreal. pp. 232-7.

Section-II: Practical Exercises

Practical Exercises

Exercise 1: Studies on climate classification

The Köppen climate classification is one of the most widely used climate classification systems. It was first published by the Russian climatologist Wladimir Köppen (1846-1940) in 1884, with several later modifications by Köppen, notably in 1918 and 1936.Later, the climatologist Rudolf Geiger introduced some changes to the classification system, which is thus sometimes called the Köppen–Geiger climate classification system.

Köppen climate classification

Köppen-Geiger climate classification map (1980-2016)

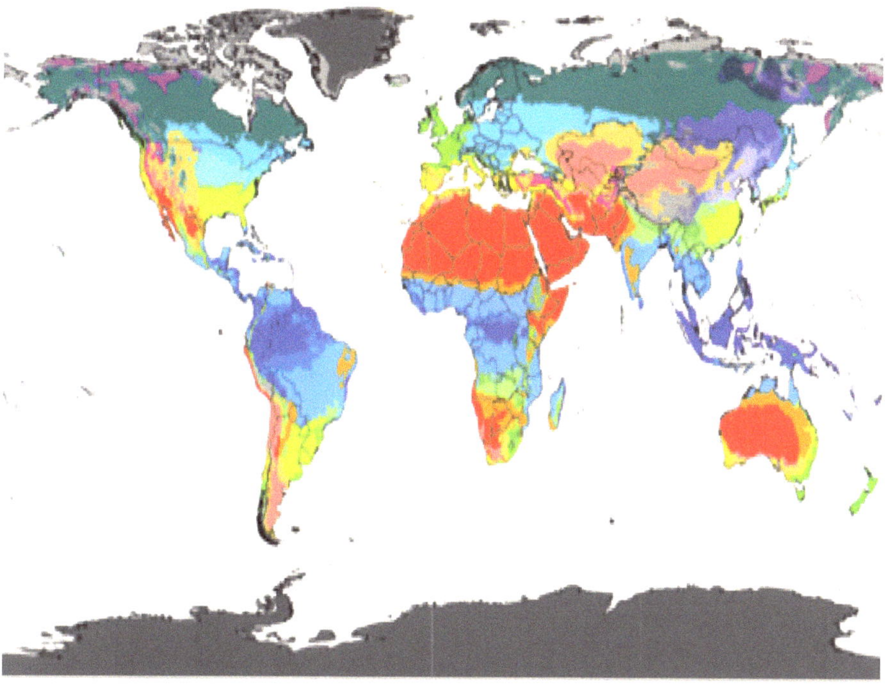

An updated Köppen–Geiger climate map

Af	BWh	Csa	Cwa	Cfa	Dsa	Dwa	Dfa	ET
Am	BWk	Csb	Cwb	Cfb	Dsb	Dwb	Dfb	EF
Aw/As	BSh	Csc	Cwc	Cfc	Dsc	Dwc	Dfc	
	BSk				Dsd	Dwd	Dfd	

The Köppen climate classification divides climates into five main climate groups, with each group being divided based on seasonal precipitation and temperature patterns. The five main groups are *A* (tropical), *B* (dry), *C* (temperate), *D* (continental), and *E* (polar). Each group and subgroup is represented by a letter. All climates are assigned a main group (the first letter). All climates except for those in the *E* group are assigned a seasonal precipitation subgroup (the second letter). For example, *Af* indicates a tropical rainforest climate. The system assigns a temperature subgroup for all groups other than those in the *A* group, indicated by the third letter for climates in *B*, *C*, and *D*, and the second letter for climates in *E*. For example, *Cfb* indicates an oceanic climate with warm summers as indicated by the ending *b*. Climates are classified based on specific criteria unique to each climate type.

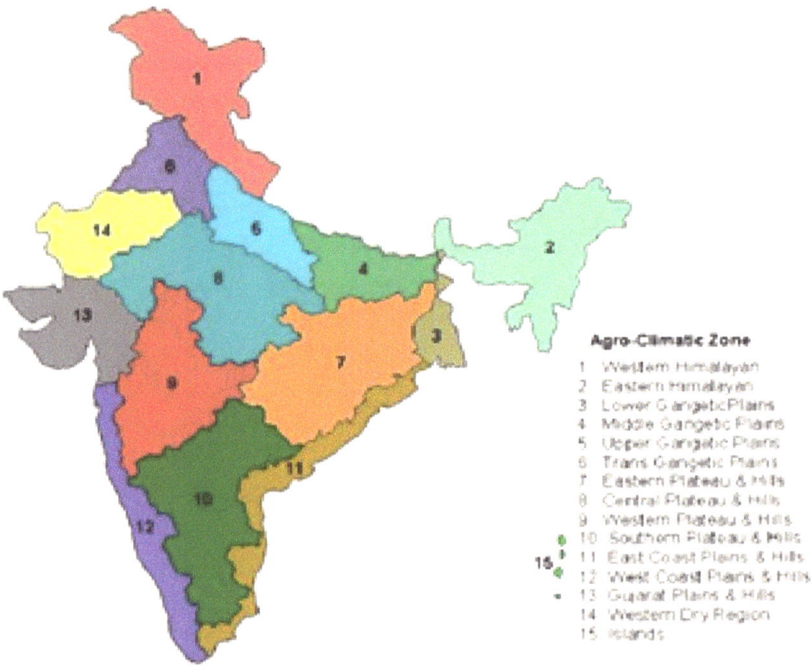

Agro-climatic zones of India map

Köppen designed the system based on his experience as a botanist, so the main climate groups are based on the different variety of vegetation that grows in climates belonging to each group. In addition to identifying climates, the system can be used to analyze ecosystem conditions and identify the main types of vegetation within climates. Due to its link with the plant life of a region, the system is useful in predicting future changes in plant life within a region.

The Köppen climate classification system has been further modified, within the Trewartha climate classification system in the middle 1960s (revised in 1980). The Trewartha system sought to create a more refined middle latitude climate zone, which was one of the criticisms of the Köppen system (the C climate group was too broad).

Exercise 2: Studies on rainfall pattern in rainfed areas of India

India has a tropical monsoonal climate since the larger part of the nation lies within the trophies, and the climate is influenced by the monsoons. The summer season in India falls from April to June and the winter season falls between November to mid-March and summer season from April to June.

Precipitation in India is extremely irregular over a period of time in a year. The western coasts (Western Ghats) and North East India receives precipitation of nearly 400 cm. the precipitation is less than 60 cm in Western Rajasthan and adjacent areas of Punjab, Gujarat, and Haryana. Likewise, rainfall is little in the insides of the Deccan Plateau and eastern parts of Western Ghats. Then, Leh is also an area of low precipitation.

Northern India region remains arid, dusty, and hostile during the summer months. The nature of monsoon is unpredictable where certain regions experience heavy precipitations the others area experience scarcity and still others get flooded. The location of the mountain ranges and direction of the rain-bearing winds are the two major elements that decide the climate of India. There are five Rainfall zones in India.

1. Extreme precipitation regions

Northeastern regions of India experiences 400cm of rainfall especially the areas like Assam, Meghalaya, Arunachal Pradesh and hilly tracts of the Western Ghats. Substantial rainy evergreen forests propagate in these areas. The wettest area in India is located in Meghalaya.

2. Heavy precipitation regions

The regions experiencing 200-300cm rainfall is belonging to this zone. The areas like Western Ghats regions, Nagaland, Manipur, Tripura and West Bengal. Wet evergreen forests are seen in these areas and the major crops comprise rice, jute, and maize.

3. Moderate precipitation regions

Areas which experiencing with 100 to 200 cm of rainfall embrace main parts of West Bengal, Bihar, Orissa, Madhya Pradesh, Andhra Pradesh, and parts of other states. Wet Deciduous forest forms the distinctive natural vegetation.

4. Scanty precipitation regions

Areas having 50 to 100 cm of rainfall comprise parts of Maharashtra, Gujarat, Karnataka, Tamil Nadu, Andhra Pradesh, Madhya Pradesh, Punjab, Haryana and Western Uttar Pradesh.

5. Desert and semi-desert regions

These are the areas having below 50 cm of rainfall contain desert and semi-desert parts of Rajasthan, Gujarat and adjacent areas. Deserts remain almost fallow land. Rajasthan and North-West Kashmir receive less than 20cm rainfall.

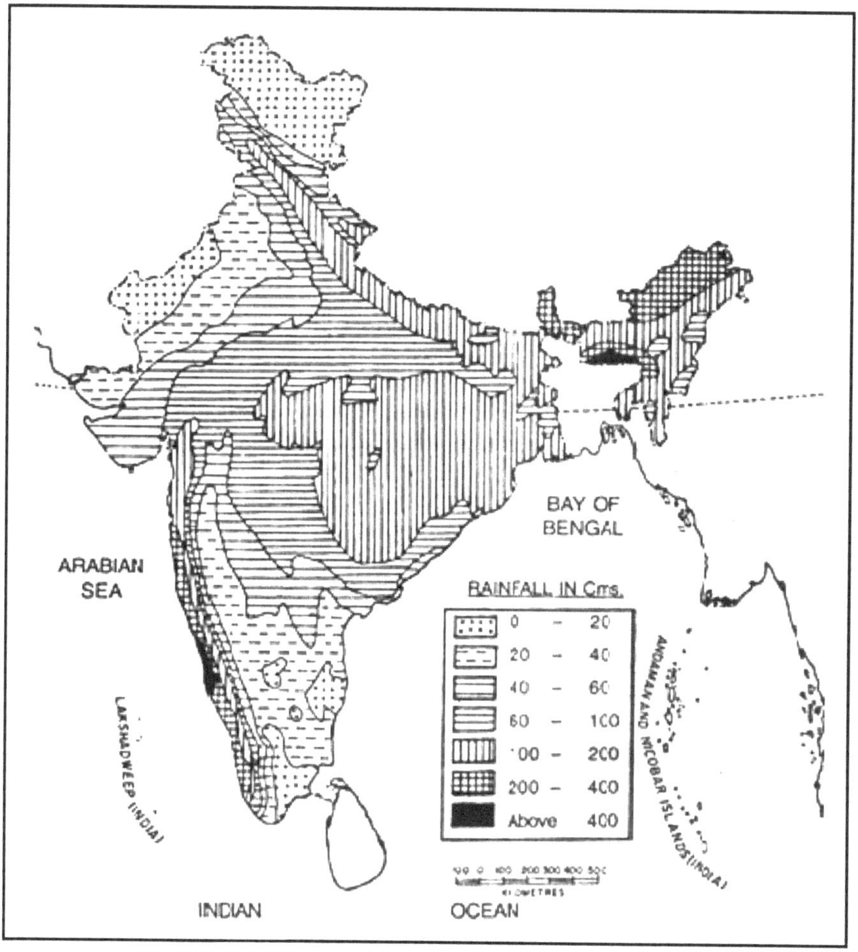

Fig. Rainfall pattern in India

Exercise 3: Status of rainfed agriculture in India

India ranks first in rainfed agriculture globally in both area (86 Mha) and the value of produce. Rainfed regions in India contribute substantially toward food grain production including 44% of rice, 87% of coarse cereals (sorghum (Sorghum bicolor), pearl millet (*Pennisetum glaucum*), maize (*Zea mays*)), and 85% of food legumes, 72% of oilseeds, 65% of cotton, and 90% of minor millets. Overall, the rainfed areas produce 40% of the food grains, support two-thirds of the livestock population, and are critical to food security, equity, and sustainability. India is home to 18% of world's population, 15% of the world livestock, 4.2% of fresh water resources, 1% of forests, and 0.5% of pasture land, but only has 2.3% of the geographical area. India is home to 25% of the world's hungry population of more than 1 billion along with an estimated 43% of children malnourished under the age of five. The net sown area in India has remained constant for several years at 141 Mha, but the human and livestock populations have been steadily increasing. Though the Indian population increased from 361 million in 1951 to 1140 million in 2011 and 1125 million in 2019, tripling over 60 years, the food-grain production has more than quadrupled, but the yield gains are largely from the irrigated agroecosystems. Notwithstanding the increase in average productivity from 0.6 Mg ha^{-1} in the 1980s to 1.1 Mg ha^{-1} at the present time, large yield gaps exist for rainfed crops in the semiarid regions. Even after realizing the full irrigation potential, nearly 40% of the net sown area of 141 Mha will remain totally rainfed. The per capita availability of land has fallen drastically from 2.4 ha in 1951 to about 0.32 ha in 2001; and it is projected to decline further to 0.09 ha by 2050. Increasing productivity of rainfed cropping systems is of critical importance to meet the food demands of an ever-increasing population in India. The potential productivity of maize in high rainfall regions under rainfed condition is 8.0 Mg ha^{-1} vis-a-vis the national average yield of 2.1 Mg ha^{-1}, indicating an unbridged yield gap of 6 Mg ha-^{1}. Large yield gaps exist in other crops as well which are primarily grown under rainfed conditions. Recommended management practices (RMPs) such as improved cultivars, site specific nutrient management (precision agriculture), and water harvesting and recycling can potentially increase the yields in several crops up to 6 Mg ha^{-1}, indicating the large realizable potential under rainfed conditions. There are many districts in India where the actual yields are much lower than the national average, and there is enormous potential for improvement.

It is estimated that "more than half of agricultural sown land, i.e. 53% is dependent on rainfed irrigation. 53% (73.83 million hectares) of the sown agricultural area (total sown area for 2012-13 was 139.93 million hectares) in the country is dependent on rainfed irrigation as of 2012-13.

Rainfed irrigation is dependent on rain water and is not supplemented by water from any other source.

Rest 47% (66.1 million hectares) of the sown area is irrigated by ground water sources such as wells and tube-wells, and surface water sources such as canals and tanks, as of 2012-13 [Net irrigated area]. This area constituted 18% (20.9 million hectares) of the sown area in 1950-51."4 See Figure below:

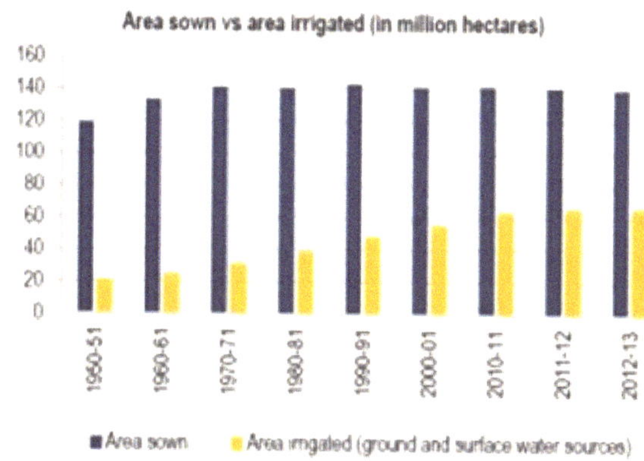

It is mostly the North Eastern states that depend on rainfall while Northern states depend on ground and surface water. "The northern states of Punjab (99.5%), Haryana (90%), and Uttar Pradesh (76%) have the highest proportion of sown area which is irrigated by ground water and surface water sources.

In contrast, less than 25% the sown area in the north-eastern states of Manipur, Meghalaya, Mizoram, and Nagaland is irrigated by ground and surface water sources. This implies that these states rely more on rain-fed irrigation.

The central states such as Maharashtra (17%) and Chhattisgarh (26%), and southern states such as Kerala (22%) and Karnataka (27%) also have lower proportions of sown area under irrigation.

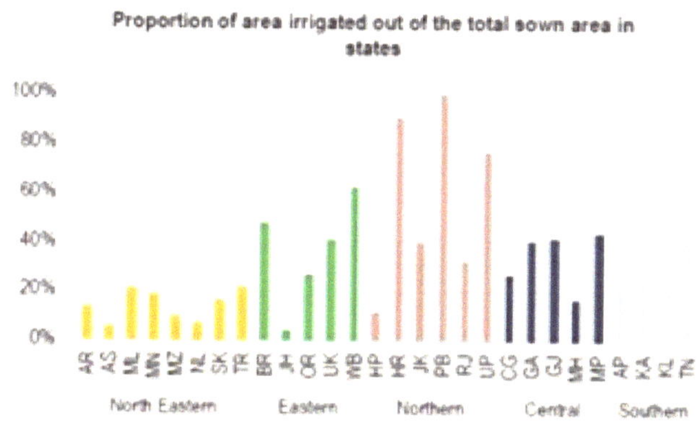

Exercise 4: Study of cropping systems for dryland areas

Cropping system is an important component of a farming system. It represents cropping patterns used on a farm and their interaction with farm resources, other farm enterprises and available technology which determine their makeup. Cropping pattern means the proportion of area under various crops at a point of time in a unit area. It indicates the yearly sequence and spatial arrangement of crops and fallow in an area. Crop sequence and crop rotation are generally used synonymously. Crop rotation refers to recurrent succession of crops on the same piece of land either in a year or over a longer period of time. Component crops are so chosen that soil health is not impaired.

Types of cropping systems

Depending on the resources and technology available , different types of cropping systems are adopted on farms.

Monocropping

Monocropping or monoculture refers to growing of only one crop on a piece of land year after year. Under rainfed condition ground nut or cotton or sorghum are grown year after year due to limitation of rainfall. Mono cropping is practiced under normal and below normal rainfall conditions by adopting improved methods of moisture conservation practices and short and medium duration varieties of the crops, either in kharif or Rabi season.

Monocropping in normal and below normal rainfall years.

Kharif crops – fallow

Fallow- rabi crops (on conserved moisture)

Multiple cropping

Growing two or more crops on the same piece of land in one calendar year is known as multiple cropping. It is the intensification of cropping in time and space dimension i.e. more number of crops within year and more number of crops on the same piece of land at any given period. It includes double cropping, intercropping, mixed cropping and sequence cropping.

Use of seeds of drought resistant, early, quick growing and high yielding varieties of the crops with improved agronomic management and cropping system depending upon the rainfall pattern bring about definite improvement in dryland crop yields. Pearlmillet, mungbean, black gram, cowpea and clusterbean during rainy season and chickpea, mustard and taramira during Rabi season were identified to be the promising crops for the dryland areas of the region.

Double cropping in above normal rainfall years:

- Pearlmillet- chickpea
- Mungbean-Mustard
- Cowpea-mustard
- Pearlmillet + Cowpea- Mustard/Chickpea

Intercropping

- Pearlmillet + Mungbean
- Pearlmillet + clusterbean

For post rainy season crops grown on conserved soil moisture, it is the available soil moisture in the soil profile at sowing time that dictates the choice of crops.

Mustard (RH-30, Varuna, RH-819 and Lakshmi) and Gram (H-208 and C-235) were identified most suitable crops for planting on well conserved soil moisture.

i) Under limited moisture availability i.e. less than 300 mm / m the response of grain yield of wheat, barley, gram, mustard and taramira to water is linear. In case of taramira this response is negative as the crop lodges due to luxuriant growth when more moisture is available.

ii) For each mm of increased available water to the crops, wheat, barley, gram and mustard yield increases but in taramira with the increasing availability of water beyond 150 mm per meter yield decreases.

iii) Wheat crop needs more than 250 mm m^{-1} water in the soil profile for obtaining satishfactory yield. The requirement of barley is about 200 mm m^{-1} and that of gram is 150-200 mm m^{-1} . For soil moisture between 125-175 mm m^{-1} mustard can be raised successfully and for low moisture (<125 mm m^{-1}) in soil profile, taramira is the only crop that can be successfully grown.

Potential cropping systems based on rainfall and soil characters

Rainfall(mm)	**Soil type**	**Suggested cropping system**
350-600	Alfisols,vertisols	Single rainy season cropping sorghum / maize/ soybean
350-600	Deep aridisols,Entisols	Single cropping sorghum / maize/ soybean in kharif / rabi
350-600	Deep vertisols	Single post rainy season cropping sorghum
600-750	Alfisols, entisols	Intercropping Sorghum + Pigeonpea Cotton + Blackgram
750-900	Entisols, deep vertisols, deep alfisols, inceptisols	Double cropping with monitoring Maize – safflower, Soybean – chick pea, Groundnut – horsegram
>900	As above	Assured double cropping Maize – chick pea Soybean - safflower

Exercise 5: Studies on mulches and anti-transpirants

Various types of mulches are as under

Mulching: With respect to dryland mulch is any material placed at soil surface with a view to conserve the soil water. These may be various types such as - Straw or residue mulch, soil mulch, plastic mulch, vertical mulch and chemical mulch.

a) Straw mulching can help in conserving soil moisture in following ways -

1. By checking loss of water from soil through process of evaporation because it reduces heating of soil from radiation and reducing wind speeds near soils surface.
2. By reducing runoff and increasing infiltration of water.
3. By checking weed growth.

b) Soil mulch: Surface mulch of dry soil 5-8 cm deep by obstructing the rise of water to the surface through capillary action effectively reduced loss of water as compared with effectively reduced loss of water as compared with a soil having an undisturbed surface. Following kinds experiment and conclusion, a soil mulch for moisture conservation became standard procedure in dry regions and dicta such as two cultivation can repoace one irrigation, became very popular.

c) Chemical Mulch: It has been found that Hexaderunol, a long chain alcohol mixed with surface 1/4 inch of the soil reduced evaporation by 43%, this material which is resistant to microbial activity remained effective for more than a year. The surface layer of treated soil dried out more rapidly that that of untreated soil, creating a diffusional barrier to evaporation.

d) Vertical mulch: Vertical muching is a recent innovation which is found to be helpful in situ conservation of moisture for increased crop yields. This has been specially found to be suitable to black soils of Decan plateau whose intake rates are very low. It consists of jowar stubbles keep in trenches of 40 cm deep, 15 cm wide protruding 10 cm above ground level. Such trenches spaced at 4-5 meters increased crop yields by 400-500% in drought years and 40-50% in normal years ever control.

Plants transpire water vapours continuously from all above ground parts particularly through leaves. This process of evaporation of water from the aerial parts of plants is termed as transpiration. Approximately 99 percent of the water taken by the plants roots is transpired to the atmosphere. Transpiration occurs through different types of apertures such as cuticles, lenticels and stomata. Among these apertures stomata accounts for 90-97 per cent of transpiration. Transpiration is considered as a necessary evil. Reduction in transpiration may help in maintaining favorable water balance in dry farming. Any material that is applied to plant surfaces with the aim of reducing or inhibiting water loss from plant surface is called antitranspirants.

Scope of Antitranspirants

- Under dryland area to reduce water losses through transpiration.
- In costly irrigation facilities for extending the irrigation intervals.
- For reducing transplanting shock of nursery plants.

Types of anti-transpirants: Based on mode of action, anti-transpirants are of four types

Stomata closing type: Most of the transpiration occurs through stomata on the leaf surface. Spray material used for various purposes such as certain fungicide (phenyl mercuric acetate, herbicide (atrazine) and metabolic inhibitors have been found to cause the closure of stomata and thereby reduction in transpiration. The effectiveness of an antitranspirants is also depends on the coverage of lower surface of leaves, interaction of the antitranspirants with external environments, surface anatomy of leaves and rate of formation and growth of new leaves.

Film forming type: Foliar spray of waxy or plastic emulsions such as mobileaf, hexadecanol and silicone produce an external physical barrier outside the stomatal opening to retard the escape of water vapour through stomatal opening. The film so formed should have more resistance to the passage of water than to that of carbon dioxide. Film type antitranspirants which provide selective type of permeability barriers to water vapours and carbon dioxide diffusion in the required directions have not yet been found so far.

Reflectance Type: White reflecting materials such as whitewash or kaoline spray form a coating on the leaves and increase the leaf reflectance (albedo). By reflecting the large amount of radiation, they reduce leaf temperature and vapour pressure gradient from leaf to atmosphere and thus reduce transpiration. Application of 5% kaolin spray has been found to reduce transpiration losses markedly.

Growth retardants: Foliar application of chemicals such as cycocel reduces shoot growth, increases root growth and induces stomatal closer. Thus, the application of such chemical helps in improving the water status in the plants and soil.

Limitations of antitranspirants

- May reduce the rate of photosynthesis
- May increase the leaf temperature by reducing evaporative cooling

- Interaction of climatic factors with antitranspirants reduces their effectiveness for longer duration
- Sometimes marginal cost is more than marginal returns.
- May produce toxic effects on leaves.

Demonstration of transpiration phenomenon

Loss of water in the form of water vapour from the aerial parts of plant is known as transpiration. The water vapour can be seen in the form of water droplets if a transpiring plant covered thoroughly

Material required: A well watered potted plant, Beljar, rubber sheet, grease, or Vaseline

Procedure

Take a small well watered plant. Cover the external soil surface of pot and its soil thoroughly with polythene bag to prevent direct evaporation of water. Invert a dry bell jar over it. Seal the edges of bell jar with Vaseline and place it in sunlight for few hours. After some time the bell jar becomes misty and its inner walls contains drops of water that may flow down the sides of bell jar. Weigh the pot. Weight of the pot is less as compared to its initial weight this reduced weight is due loss of water in transpiration.

Precaution

- The bell jar should be air tight at rim base by putting grease or Vaseline.
- The pot should be well watered before starting the experiment
- The bell jar should be made up of glass to penetrate light for photosynthesis

The pot surface and soil must be covered carefully to avoid evaporation losses.

Exercise 6: Seed treatment, seed germination and crop establishment in relation to soil moisture contents

Seed Treatment

Seed treatment refers to the application of fungicide, insecticide, or a combination of both, to seeds so as to disinfect and disinfest them seed-borne or soil-borne pathogenic organism and storage insects. It also refers to the subjecting of seeds to solar energy exposure, immersion in conditioned water, etc. The seed treatment is done to achieve the following benefits.

1. Benefits of seed treatment

1. Prevention of Spread of Plant Diseases .The disease from treatment standpoint may be conveniently grouped under two types-

a) Systemic Disease : That infect the seed during the harvest or storage period resulting in infection of seed , E. g Bunt or stinking smut of wheat, Helminthosporium blight of barley, loose and covered smut of oats; head and kernel smuts of rye , smuts of millet . Appropriate seed treatment is significantly effective in controlling these diseases. That infects seed during the flowering stage to become established within the seed and from there within the resulting plant. Such diseases include loose smuts of wheat. Treatment with systemic fungicides, E.g Vitavax has been found effective.

b) Non-systemic Disease : Diseases that infect seed during the harvest or storage period. Such diseases includes Helminthosporium blight, blotches or blight of barley, oats, rice , rye , sorghum, wheat and Fusarium. These diseases can be effectively controlled by appropriate seed treatment.

2. Seed treatment

Protects seed from seed rot and seedling blights. Seed treatment, by its protective coating around the seed, acts as a barrier once the seed is planted to ward off attack by both seed-borne and soil-borne organisms. These organisms affect, all crop seeds and the degree of attack depends upon a number of factors of particular importance are the organisms. Pythium spp, Rhizoctonia and Sclerotium that are present in all soils. They may rot the seed before germination gets well started, or they may kill the seedling before it emerges, or so affect it that it dies after emergence or supervives only as a weakened plant. The responses to protective treatment varies with the kind of crop seed, the vigour of the particular seed, the amount of mechanical injury to seeds, conditions of seed surface and adversity of planting conditions. The fungicide treatment compensates by protecting these cracks and abrasions from entrance of fungi.

3. Improves germination

Seed treatment often improves the standard of germination through the control of seed surface flora, though normally not considered pathogenic; this may infect the seed following moist harvesting and storage conditions. In the germination test it may smother the seed before it has a chance to germinate.

4. Provides protection from storage insects

The protection of seed from insect damage during storage is of increasing importance with the trend towards processing, treating and unit packaging of seeds at harvest time. For complete protection it is necessary to treat seed with insecticide also. Insecticides are more needed in warm storage than cool storage.

5. Controlling soil insects

This can be done through combination treatment – the process of addition of an insecticide with fungicide for the added protection of the seed and seedling against certain soil insects, such as wire worm and the seed corn maggot. In contrast to storage insect protection, it is a means of giving limited protection to the seed and seedling until it becomes resistant to attack or can survive limited attack. It is not a means of disinfecting the soil.

Seed germination

Seed germination means the resumption of growth by embryo & development of a young seedling from the seed. Germination is an activation of dormant embryo to give rise to radical (root development) and plumule (stem development). Germination is the awakening of the dormant embryo. The process by which the dormant embryo wakes up & begins to grow is known as Germination. Seed Emergence means actually coming above and out of the soil surface by the seedling.

Changes during germination

1. Swelling of seed due to imbibitions of water by osmosis.
2. Initiation of physiological activities such as respiration & secretion of enzyme.
3. Digestion of stored food by enzymes.
4. Translocation & assimilation of soluble food.

When seed is placed in soil gets favorable conditions, radical grows vigorously & comes out through micro Pyle & fixes seed in the soil. Then either hypo or epicotyls begins to grow.

Factors affecting the germination

External factors

1. Moisture: It enables the resumption of physiological activities, swelling of seed due to absorption of moisture & causes bursting of seed coat & softening the tissue due to which embryo awakes & resumes its growth.
2. Temperature: A suitable temperature is necessary for proper germination. Germination does not take place beyond certain minimum & maximum temperature i.e. 0p C & above 50p C. Optimum temperature range for satisfactory germination of seed is 25 to 30p C.
3. Oxygen: It is essential during germination for respiration & other physiological activities which are vigorous during the process.
4. Light: It is not considered as essential for germination & it takes place without light. The seedlings grow more vigorously during darkness rather in light. However, for survival of germinating seedling, light is quite essential.
5. Substratum: It is the medium used for germinating seeds. In the laboratory, it may be absorbent paper (blotting paper, towel or tissue paper), soil & sand. Substratum absorbs water & supplies to the germinating seeds. It should be free from toxic substances & should not act as medium for growth of micro-organisms.

Following factors are to be taken into consideration for seed germination

1. Dormancy period - A seed requires to undergo a period of dormancy after its formation. During this period the seed 'rests' and may refuse to germinate properly. This period varies from seed to seed. Seeds of water melon or red gourd do not have a dormany period and may germinate easily immediately after their formation in the fruit. Others require dormancy period ranging from a few weeks to a few months.
2. Viability of a seed - It is exactly like expiry date/period of a commercial product. It must not be confused with the dormancy or 'resting period' after the resting period a seed becomes ready to germinate. But this readiness is not forever. It literally expires after certain period. After that seed will not germinate. This 'ready to germinate' period is the viability period of a seed. It may vary from a few years to many centuries. Seeds of Lotus are reported to have germinated even after eight hundred years! In that case their viability period is at least that much if not more.

External factors

1. Soil moisture or water: This is required to trigger the mechanism of germination. In the absence of moisture the seeds cannot germinate; but when it is available it is imbibed by the seed coat and the enzymes in side become active and functional. The amount of water does not matter in the initial stages; but later on it becomes critical.
2. Soil Texture: Does not matter during germination. Seeds can even germinate without soil on the piece of a moist blotting paper.
3. Soil pH: Should too acidic. pH 5 to 8 is alright.
4. Soil temperature: It should be on the warmer side. Lesser the temp. lesser the rate of germination.
5. Light: It does not affect the process of germination; but some seeds do germinate better in the absence of light. They like darkness for germination it seems.

Exercise 7: Watershed: characteristics, delineation and visit to dryland research stations and watershed projects

Characteristics of watershed

Each watershed shows distinct characteristics, which are so much variable that no two watersheds are identical. All these characteristics affect the disposal of water. Several characteristics namely: size, shape, slope, drainage, vegetation, geology, soil, climate, land use etc. are considered as important.

Size

The size of a watershed forms a basis for further classification into different categories as :

a) Sub watershed (100-500 sq. km)

b) Milli watershed (10-100 sq. km)

c) Micro watershed (1-10 sq. km)

d) Mini watershed (less than 1 sq. km)

The size helps in computing many parameters like precipitation received, retained and drained off. Large the watershed, more the heterogeneity of the other characters.

Shape

Watersheds differ in their shape based on morphometric parameters like geology and structure. The general shapes are elongated, triangular, circular etc. shapes determines the length with ratio which affects the run off characteristics like run off time.

Physiography

Type of land, its attitude and physical disposition immensely speak about a watershed as to the climate and planning the activities in greening. For example a hilly tract could be useful mainly for trees and plains of populated area could be utilized only for crops.

Slopes

It controls the rain fall distribution and movement, land utilization and watershed behaviour. The degree of slope affects the velocity of overland flow and run off, infiltration rate and thus soil transportation.

Climate

Meteorological parameters like precipitation, temperature, wind velocity, humidity and evaporation decide and quantitative approach for arriving at water availability in a watershed. Climate is a determining factor for management etc. It determines the flow characteristics and thus erosion behaviour.

Vegetation

Detailed information on vegetation helps in choosing tye, made and manner of greening the watershed. Information on local species give a sure guide for selecting plants and crops. It confirms what greenery can be grown and where. With care soil capabilities could be analysed, compared and profitably confirmed for management.

Geology and soils

Rocks and their structures control the formation of a watershed itself their nature determines size, shape, physiography, drainage and ground water conditions, soil parameters as to depth, nature, moisture and fertility determine crops. Rock and soils together influence water storage, movement and infiltration.

Hydrology

Availability, quantity, and distribution of surface water is basic to the final goal of growing greenery in a watershed. Hydrology parameters help in quantification of water available, utilized and additional exploited resources. They determine the location and design of conservation structures.

Socio-economics

Statistics on people, and their health, hygiene, wants, wishes, cattle and farming practices and share of participation are important in managing watershed.

Hydrogeology

The demand for ground water is ever on the increase. As such, the appreciation of ground water resources for determining their further availability in the context of conjunctive use of water.

Delineation of watershed

Hydrologically, watershed is an area from which the runoff flows to a common point on the drainage system. Every stream, tributary, or river has an associated watershed, and small watersheds aggregate together to become larger watersheds. Water travels from headwater to the downward location and meets with similar strength of stream, and then it forms one order higher stream. The stream order is a measure of the degree of stream branching within a watershed. Each length of stream is indicated by its order (for example, 1rst-order, second- order, etc.).

The start or headwaters of a stream, with no other streams flowing into it, is called the 1rst-order stream. First-order streams flow together to form a second-order stream. Second-order streams flow into a third-order stream and so on. Stream order describes the relative location of the reach in the watershed. Identifying stream order is useful to understand amount of water availability in reach and its quality; and also used as criteria to divide larger watershed into smaller unit. Moreover, criteria for selecting watershed size also depend on the objectives of the development and terrain slope. A large watershed can be managed in plain valley areas or where forest or pasture development is the main objective. In hilly areas or where intensive agriculture development is planned, the size of watershed relatively preferred is small.

Visit to important watersheds and dryland institutes

The impact and effectiveness of the watershed initiatives such as Sukhomagri and Bunga have clearly shown that it is possible embed ecologically sound rainfed agriculture in economically sustainable development. The other good examples are made at Mir (Udhampur, J & K); Algai Pandiyapuram (Tirunelveli, T.N.), Nartora (Raipur, M.P.), Jhanwar (Jodhpur), Gursutinala (Palamau, Bihar), Padal Singhi (Beed, Maharastra and Dapoli (Mahrastra).

1. Bunga watershed: The project having an area of 463 ha is located at village Bunga in district Ambala.The average annual rainfall is 1115 mm. The soil is loamy sand and sandy loam with gravels on the surface.
2. Bajar-Ganiyar watershed: The project is having an area of 349 ha in Bajar and 746 ha in Ganiyar and is situated in district Mahendergarh. The average annual rainfall is 640 mm.
3. Apart of Shivalik which forms foothills of Himalayan ranges also lies in the district of Panchkula, Ambala and Yamunanagar of Haryana state. This area, 1.92 lakh ha forms part of five watersheds namely Sirsa, Ghaggar, Dangri, Markanda and Yamuna. This is further divided into 16 sub watersheds and 162 micro watersheds. The area was taken up of development under IWDP in the year 1990 and lasted upto1999. There after it continues as Phase II. The main objective of the project is to restore the productive potential of the project area by developing osil-plant-water resources of the watershed to rectify the man made ecological imbalance by conserving the natural resources on arable and non arable lands.

Visit to important dryland research stations can also be undertaken for updating the knowledge of the students. Important institutes are CRIDA, ICRISAT, CAZRI and AICRP in drylands in important agricultural universities.

Exercise 8: Studies on agronomic measures for soil and moisture conservation

Various method of moisture conservation in dryland areas are by storing in the root zone, by checking loss of water by evaporation and by water harvesting. Rainfall water can be conserved in following ways:

(A) By storing more of rainfall in root zone.

1. Tillage,
2. Contour farming of bunding,
3. Vertical mulching,
4. Subsurface barrier,
5. Addition of pond sediments of organic matter
6. Addition of gypsum

(B) By checking loss of water through evapotranspiration

1. Mulching,
2. Weed control
3. Use of antitranspirants,
4. Crop thinning

(C) By water harvesting

1. Water harvesting *in situ.*
    a) Inter row water harvesting
    b) Inter plot water harvesting
2. Water harvesting for recycling

Elaborate above agronomic measures in your own words.

Exercise 9: Study of implements used in dryland agriculture

The need to have improved implements for seeding, fertilization and interculture operations was long being felt in dryland areas.

Ridger seeder (Tractor drawn)

Bajra the principle kharif dryland crop suffers at germination stage due to problem of soil crusting/seed burying by rains after sowing. Ridge sowing of the crop could help to get rid of this problem. Also there was a definite problem to get proper germination of rabi oilseed crops raised on conserved moisture or as a second crop after short duration kharif pulse crop due to receding condition of soil moisture in the profile, ridger seeder is quite suitable implement to mitigate this problem. In kharif the ridger seeder place the seeds on top of the ridges in 30/60 cm paired row system and in rabi place the seeds deep in the moist zone and ensures proper coverage with moist soil.

Construction details

It consist of a tool bar holder for three point hitch mounting to hold a 2 inches square tool bar, 9 feet in length. Three ridgers provided on tool bar, form the ridging assembly. The wings of the ridgers are adjustable to form ridges and furrows of various dimensions. The 9 inch length of the tubing mounted to the ridger frame is shaped to direct the flow of fertilizer to the bottom of the furrow. Hoe type furrow openers for seed placement are mounted to the tool bar so that the openers are behind the fertilizer placing device and seed is placed above the fertilizer. The depth of seedling can be varied by adjusting the position of the hoe openers relative to the tool bar and ridger assembly. The furrow openers can also be adjusted to sow the seed in furrows or on the side of the ridges. A lugged gauge wheel is provided to regulate the depth of furrows and to drive the seed and fertilizer attachment.

A seed box fabricated from galvanized metal sheet is provided. Fluted roller type seed metering system to meter out the seeds ranging from chickpea to rapeseed is provided. A north western fertilizer attachment was mounted ahead of the seed box.

Multipurpose seed drill cum ridger seeder (tractor drawn)

Farmers owing tractor in Haryana has drylands and irrigated lands sow wheat, cotton and other irrigated crops in his irrigated areas. The ridger seeder as such was not acceptable to the farmer due to its inability to sow wheat. To increase the versatility of ridger seeder as multipurpose seed drill cum ridger seeder was developed. Multipurpose seed drill cum ridger seeder has detachable arrangement which can be used to make the machinery as seed drill as well as ridger seeder.

Recently, arrangement for sowing of intercropping system was also included in multipurpose seed drill cum ridger seeder.

Zero-till machine

Constructional details

It consists of a rectangular tool bar 200 cm X 60 cm. Seed and fertilizer boxes fabricated from galvanized metal sheet provided with fluted rollers each to meter out seed and fertilizer is provided. Nine combined furrow openers are provided to make it nine row seed drill. Two big clamps to fix two ridge bottoms on the main front tool bar are provided to make it ridger seeder. This implement was tested to sow Bajra, cotton, moong,guar in kharif and mustard , wheat, gram in Rabi and found quite promising. The wheat was sown at row spacing of 17 cm. The germination was found very well.

Ridger seeder (Bullock drawn)

Tractor drawn machinery helps in speedy work and timeliness of operation, but tractor power is not within easy reach of small farmers. A simple attachment on country plough to enable it to sow the seeds on ridge as well as in furrow was designed. This implement was vigorously tested and some shortcomings were noted.

Interculture implements

Moisture is a limiting factor in raising a good crop. Weeds can not be allowed to sustain as these will compete with plants for nutrients as well as for moisture. The rapidity of interculture and weeding operation plays a vital role as timeliness of weeding is must. It was found that by breaking the soil crust and forming shallow soil mulch in field helps in maximum moisture conservation.

Bullock drawn blade hoe

It consists of a frame fabricated from mild steel flat size (40 X 10mm) in such a way that different size of blade can be fitted in the frame. Various blades of dimension 5 mm thick and 100 mm wide of 250 to 450 mm length fitted in the frame such that it makes a cutting angle of 15- 20 degree with the soil surface. The frame is provided with a 80 x 50 mm beam made of seasonal hard wood. The length of the beam is 2.5 metre.

Wheel hand hoe

It consists of a mild steel wheel having 400 m dia. The wheel is provided with eight spokes of 6 mm thick rod. Wheel hub with cast iron bush and mild steel axle is also provided A mild steel frame of variables sizes 150mm to 250 mm are mounted on the wheel. Two handless made ¾ inches dia 16 gauge steel pipe provided with 125 mm grips are attached on the frame. This implement tested and found labour saving having 1man/acre/day coverage.

Testing and evaluation of Modified Ridger seeder for Pearlmillet andRaya sowing

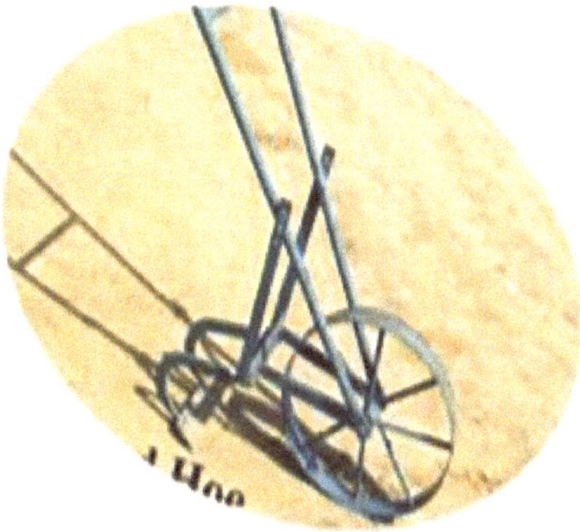

Wheel hand hoe

Multi purpose seed cum fertilizer drill

Bed planter

Zero till drill

Laser land leveler

Exercise 10: Study on methods for soil moisture determination

Soil is a heterogeneous mass and consists of three phases, viz., solid, liquid and gaseous phase which serves as a water reservoir or bank. Determination of soil moisture content is needed to help in the estimation of

a) Available water in the root zone,

b) Scheduling of irrigation,

c) Soil water potential and

d) Changes in physical and chemical properties of soil due to changes in water content.

Broadly, the methods used for determination of the soil water content are classified as Direct methods and Indirect methods.

i) Direct methods: In these methods the techniques involve either the direct measurement of water present in a soil sample or the loss of water from a soil sample. These methods involve the removal of water from the soil samples by evaporation, leaching or chemical reaction, where soil and water are separated, and the amount of water removed is measured or the change in weight of the soil due to moisture loss is estimated. Gravimetric methods fall in this category.

ii) Indirect methods: These methods involve measurement of soil properties or determination of properties of some object placed in the soil, which are affected by the soil water content. Indirect methods are preferred over direct methods because they are non-destructive, less time consuming and permit more frequent observations, however, these methods require an initial calibration.

The classification of various soil moisture determination methods is as below:

i) Direct method i.e Gravimetric method

ii) Indirect methods

Direct methods of soil moisture determination (gravimetric method)

(A) Weight basis

Materials: Sampling screw auger, moisture box, weighing balance and drying oven.

Procedure

1. Take a composite sample of soil about 50 g in a moisture box and cover it immediately with its lid.

2. Cover the box with a cloth in the field to avoid heating due to insulation. Carry the samples to the laboratory.
3. Tipping off some soil near the lid, weigh the sample on a balance correct to two decimal places in gram. If field balance is available, the moist soil sample can be weighed just after sampling in the field (wet weight).
4. Dry the sample in an oven to a constant weight at 105°C. This takes about 8 hours. Record the weight of the dried sample (dry weight).
5. Weight the empty moisture box.
6. Calculate the moisture percentage on weight basis by the formula:

$$\frac{\text{Mass of water in soil sample}}{\text{Mass of dry soil sample}} \times 100$$

(B) Volume basis

Materials: Sampling tube or a core sampler, moisture box, weighing balance and drying oven.

Procedure

i) Take the soil sample with the help of core sampler of known volume (V).
ii) Follow the steps for determination of soil moisture as in weight basis.
iii) Calculate the moisture percentage by the relationship,

$$\frac{\text{Mass of water in soil sample}}{\text{Mass of dry soil sample}} \times BD \times 100$$, *Where* BD is the bulk density.

(C) Depth basis

Materials: Sampling tube or a core sampler, moisture box, weighing balance and drying oven.

Procedure

i) Take the soil sample with the help of core sampler of known volume (V) up to desired depth
ii) Follow the steps for determination of soil moisture as in weight and volumetric basis
iii) The moisture content on depth basis can be calculated by the formula

$$\frac{\text{Mass of water in soil sample}}{\text{Mass of dry soil sample}} \times BDxD \times 100$$

Where BD is the bulk density and D is the depth of soil in cm..

Observations

i) Weight of the moisture box with wet sample = ... g

ii) Weight of the moisture box after drying sample = ... g

iii) Weight of the empty moisture box = ... g

iv) Bulk density soil = ... g/cc

v) Depth of soil profile =............ cm

Exercise 11: Soil moisture determination by neutron moisture meter

This is an indirect method of soil moisture determination which uses the interaction of neutron with water molecules as a basis of monitoring soil water content in situ.

Materials: Neutron moisture meter, aluminum access tubes, balance, oven, aluminum boxes and screw auger/tube auger.

Procedure

1. Install the aluminum access tube of specific diameter with closed bottom, up to the crop rooting depth. For this, a hole of diameter equal to outer diameter of the access tube is made in the field. The access tube is inserted in it, keeping- 20-30 cm above the ground. Generally a 2-meter length access tube is installed. The gap between outer circumference of the tube and inner wall of the hole is filled with soil slurry in order to ensure proper contact of the access tube with the field soil. Cover the opening of the tube with a rubber stopper.
2. Before taking actual observation for the soil moisture, take standard count (Ns)by placing the neutron moisture meter on the top of the open access tube,keeping the instrument in ON position and timer at 4 minutes/calibration mark.
3. Lower down the probe into the access tube up to the desired soil depth. Put the instrument on measurement mode and the timer at the desired option of 4 minutes or 1 minute or 30 seconds. After the desired time record the measured count (Nm). Continue the same process to measure the count at lower depths.
4. Find out soil moisture against the count ratio (Nm/Ns) from the calibration curve prepared for this soil by measuring various count ratios and corresponding water content values at different soil depths. Obtain directly the soil moisture from the instrument in case there is built in microprocessor in the unit.

Calibration: Initially, standard counts are taken with the probe inside the field. The shield acts as a standard having moisture 50 per cent by weight. Counts are generally taken for 1 minute intervals but for precision, they should be taken for greater intervals. Observations are taken successively at 30 cm intervals after lowering the probe in the access tube. Count ratio is then computed by dividing the observed counts (Nm) by the standard counts (Ns). The Calibration curve is obtained by taking series of counts and soil samples for gravimetric estimation simultaneously at different depths.

Merits

i) It is very quick, non-destructive method.

ii) It measures the soil water content on volume basis, in the whole range from saturation to wilting point.

iii) It is not influenced by hysteresis or salts present in soil.

Limitations

i) It is an expensive technique.

ii) Surface measurements are normally unreliable.

iii) It gives overestimation of soil moisture content in the presence of high amount of organic matter, cadmium, boron etc. and requires careful handling.

Exercise 12: Critical growth stages to water stress for major field crops

Crop	Crop duration (days)	Critical stages for irrigation
Wheat (dwarf)	130-140	Crowing, flowering, jointing, milk and dough
Wheat (tall)	140-150	Tillering, heading and dough
Barley	125-135	Tillering, booting
Rice (dwarf)	130-140	Flowering, grain filling, tillering
Rice (Basmati)	140-150	Flowering, grain development
Summer maize	90-110	Tasseling and silking, and grain development
Winter maize	150-180	Tasseling, grain development
Pearlmillet	75-85	Heading
Sorghum	90-110	Flowering, seeding, grain filling
Cotton	150-180	Flowering and boll development
Chickpea	130-140	Preflowering and pod development
Pigeonpea	150-170	Flowering initiation, pod development
Moongbean	65-75	Flowering
Clusterbean	130-145	Branching and flowering
Mustard	140-150	Pre-flowering and pod development
Groundnut	110-135	Pegging, pod-setting and pod filling
Sunflower	100-120	Flowering, seed filling, late vegetative
Sugarcane	280-330	Tillering and grand growth

Water requirement (WR) of major dryland crops and fruit trees

Summer crops	WR(mm)	Winter crops	WR(mm)	Fruit crops	WR(mm)
Pearlmillet	250	Rocket plant	150	Gonda	560
Mungbean	225	Mustard	200	Ber	660
Clusterbean	275	Chickpea	250	Aonla	660
Cowpea	240	Lentil	285	Clusapple	860
Pigeonpea	450	Peas	300	Guava	1160
Maize	400	Barley	300	Pomegranate	1160
Sorghum	400	Wheat	350	Mango	1260

Exercise 13: Studies on water harvesting structures

Since time immemorial, several kinds of water harvesting structures have been in practice in our country. Recently, local experiences and precedence's were utilized for designing these structures. In order to bring cost effectiveness, efforts are also necessary to harmonize the designing of water harvesting structures based upon traditional wisdom and scientific contemplations. In fact, farm ponds, surface water tanks and percolation tanks need to be viewed as 'conservation structures in addition to sources for supplemental irrigation. In real watershed situations, ponds maybe located in conjunction with waterways some maybe of a few hundred cubic metre size, .whereas couple of structures may be of bigger size. In selected pockets, integrated catchments cum command approach should be followed. To ensure proper storage in light soils, some of the ponds can be lined with sealant materials. All large ponds have to be constructed at Government cost. All such structures are likely to recharge ground water reservoir. When located in proper co cultural hydrologic matrix, in conjunction with other factors, they also help in reducing load on ground water aquifer. It maybe advisable to work out design details of all the required structures in the first phase of the project, no matter construction of a few of these may have to be deferred may be planned to be taken up in drought years to provide employment to the rural folk. The objectives of water harvesting structures are:

- To store water for supplemental and off season irrigation;
- To act as silt detention structure;
- To recharge the ground water;
- To raise aquaculture species; and
- Recreation and allied agricultural uses.

Thus in most situations, in addition to reviving traditional water harvesting systems, other measures are also needed. Some of such typical water harvesting structures adopted in watersheds is as follows:

Minor irrigation-tanks: *Minor* irrigation tanks are constructed across the major streams with canal system for irrigation purpose by con-structing low earthen dams. A narrow gorge should be preferred for making the dam in order to keep the ratio of earthwork to storage as min-imum. The height of the dam may vary from 5 to 15 m. The tanks are provided with well-designed regular and emergency spill ways for safety against side cutting. In micro watersheds, water harvesting *bundries,* similar to small scale irrigation tanks are also recommended. By and large, the water harvesting *bundhies* are not integrated with extensive canal system.

Farm ponds: Farm ponds are typical water harvesting structures constructed by raising an embankment across the flow direction or by excavating a pit or a combination of both. Dug out ponds in light soil require lining of the sides as well as the bottom with suitable sealants. This is not the case with heavy black soil. Considering the benefit of the harvested water, different type of linings, viz., brick, concrete, HDPE, etc. may be used.

Nala bunds and percolation tanks: Nala bunds and percolation tanks are located in the nalas having permeable formations with the primary objective of recharging the ground water. A strict regulation on the silt load entering the downstream reservoirs is an additional advantage of percolation tanks. The percolation tanks encourage the digging up of wells downstream of recharged area for irrigation purposes. The percolation tanks are provided with emergency spillways for safe disposal of flow during floods.

Stop dams: *Stop* dams are permanent engineering structures constructed for raising the water level in the *nala* for the purpose of providing life saving irrigation during drought periods. These are located over flat *nalas* at narrow gorges carrying high discharge of long durations. They are created over stable foundation conditions where hard murrum or rocks are encountered. For the stop dam, a site with larger water storage capacity would be desirable.

Compartmental bunds: *Compartmental* bunds, converting the area into square/ rectan-gular parcels are useful for temporary im-pounding of water for improving Moisture status of the soil. These are made using bund formers. In medium deep black soil, they are found ad-vantageous in storing the rainfall received during the rainy season in the soil profile there by augmenting the soil moisture. for use by rabi crops. The size of the compartments may be fixed' considering the size, of the inter bunded land.

Broad bed and furrows: Broad bed and furrow system furrows. This technique is especially suited to black soils, where crops are sown on pre formed beds. This system is made before the season and is maintained year after year. The planting is done on the bed. Generally, the depth of each furrow is kept 0.15 m and the inter furrow spacing is maintained at 1.5 m.

Dead furrows: Dead furrows are laid across the land slope in rolling lands, to intercept the run off. The spacing between dead furrows varies between 2 to 5 m or 4 to 7 crop rows. This system works well in alfisols.

Land and water conservation practices

Soil and water conservation practices are the primary step of watershed management program. Conservation practices can be divided into two main categories:

1. *in-situ* and
2. *ex-situ* management.

Land and water conservation practices, those made within agricultural fields like construction of contour bunds, graded bunds, field bunds, terraces building, broad bed and furrow practice and other soil-moisture conservation practices, are known as in-situ management. These practices protect land degradation, improve soil health, and increase soil-moisture availability and groundwater recharge. Moreover, construction of check dam, farm pond, gully control structures, pits excavation across the stream channel is known as *ex-situ* management. *Ex-situ* watershed management practices reduce peak discharge in order to reclaim gully formation and harvest substantial amount of runoff, which increases groundwater recharge and irrigation potential in watersheds.

FSC
www.fsc.org
MIX
Papier | Fördert gute Waldnutzung
FSC® C083411

Zeitfracht Medien GmbH
Ferdinand-Jühlke-Straße 7
99095 Erfurt, Deutschland
produktsicherheit@kolibri360.de